Y.B. Sokolovsky, V.M. Rotkin

THEORETICAL AND TECHNICAL BASIS FOR THE OPTIMIZATION OF WIND ENERGY PLANTS

Lulu Press, Inc.
Copyright © 2017

Contact Information:
Yuly Sokolovsky - sokol1937y@gmail.com,
Vladimir Rotkin - ricentr@mail.ru

ISBN 978-1-387-33326-4

9 781387 333264

Published & printed in Lulu Press, Inc.,USA

Summary

The work contains an analytical review of modern means of wind power, based on different physical principles. Considered as widespread wind installations with a horizontal-axial configuration, a variety of vertical-axial installations with the Darrieus Rotor, with differential drag, with adjustable blades, and innovative designs for bladeless wind turbines. Also, modern methods and possibilities for scaling and optimal regulation of wind farms are considered.

Mathematical models for the interaction of airflow with blades of collinear and orthogonal plants have been developed, and methods for optimal regulation have been identified that increase the energy efficiency of installations. Calculation techniques that allow to perform the refined calculations of the plants, carry out their simulation and numerical experiments are obtained.

A number of copyright patent developments are proposed that represent promising directions for the development of wind energy: horizontal and vertical collinear installations, vertical-axial orthogonal devices with unregulated and self-regulating blades, and with optimal control systems based on author's optimization techniques.

Contents

Introduction. General questions of wind energetics.

In conditions when possibilities of traditional sources used have been limited energy saving and using of alternative sources of energy acquires big importance. Wind energy plants are one of effective energy sources.

The Sun is a source of energy which due variable as for time and square of radiation causes irregular heating of air masses. These heat differences lead to well-balanced transferring of air masses – airflow. A density of warm air is less than cold one so it raises on the equator at the height of 10 km and spread to the northern and southern directions. Because of the Earth rotation two forces of inertia which are perpendicular to the speed vector of air movement influence on air masses: centrifugal force which acts along meridional direction and causes breaking of flow and force of Koriolis which causes deviation of flow to the right side in the northern hemisphere and to the left side in the southern hemisphere relatively to initial direction of movement.

During long time wind energy (airflow) is considered as a pure ecological inexhaustible source of energy. An energy which is radiated from the Sun and transformed into kinetic energy of wind streams on the Earth corresponds to above 10^{11} GW of total power according to the expert estimation.

Growing demands of mankind in energy are satisfied in main at the expense of traditional fuel. Threat as for lack of inexhaustible source of energy and growth of dependence from imported fuel lead to activity of researches connected with the transformation of airflow in the suitable type of energy [1-10]. More than 100000 of industrial wind energy plants of different power work nowadays in the world. There are wind energy plants in dozens of countries. USA, China, Germany. Spain and India are leaders in the production of wind electrical energy. They produce 70% of all wind energy. Wind energetics takes considerable place in the energetics of a number of countries: in Denmark, it is 20%. In Portugal 13%, in Spain 14%, in Germany 9% in a total quantity of produced energy. China is the leader in the rates of development of this industry. During 4 years the power of wind energetics has been increased twice in this country. The most powerful wind electrical station is placed in Roscoe Wind Farm, Taxis USA. It unites 627 wind plants, which produce together 781.5 MW of electrical energy. Off-shore wind plants work in 12 countries of the world mainly in Europe: in Germany, Sweden, the Netherlands. The biggest wind plant Horns Rev2, which is placed on the water was built in 2009 in the Northern Sea near the coast of Denmark. It includes 91 turbines with a total power of 209.3 Mw.

One of the positive features of wind plants is this fact that they work in unison with our needs. In the most areas of the world the strongest wind blow in autumn and at the beginning of winter when a man needs in light and heat most of all. And on the contrary the times of lull - mainly in summer coincide with the periods in reducing of consumed energy (it concerns home consumption) The main obstacle in using of wind energy is economic – power of a plant remains small and share of expenditures is considerable. However, the cost price of energy permits wind plants really competes with traditional sources of energy.

Using of wind energy for producing electrical energy has a lot of advantages. When we use it ecological harmful gases which influence on climate are not produced. In comparison with traditional energy resources wind energy is endless and free. However, devices for it producing need considerable capital expenditures. There are shortages connected with non-stable winds, weather, local and climatic risks, special ecological limits.

For a better understanding of the potential of wind energy and the possibilities of its creation, it is useful to consider an approximate calculation of the wind power plant (Table 1). Wind velocity is one of the main characteristics of the airflow, which determines its energy. Measured in meters in seconds (m/s).

Table1. The preliminary calculation of main parameters for vertical-axial wind energy plant.

Parameters	Formula	Meaning
Air density kg\m2	Relative constant	1.20
Limited index of wind energy using(IWEU) of wind plant	Constant	0.593
Efficiency of generator 0..1	Relative constant	0.82
Efficiency of electronic invertor 0...1	Relative constant	0.85
Total efficiency of plant 0...1	Relative constant	0.44
Velocity of wind m\sec	Given meaning	11
Outward diameter of rotor, m	Given meaning	3.00
Height of rotor, m	Given meaning	3.00
Orthogonal section of rotor ,m	Outward diameter of rotor x Height of rotor	9.00
Volume of air expenditure through orthogonal section, m3/sec	Volume x Air velocity	99
Mass air expenditure through orthogonal section, kg/sec	Volume x Air density	119
Power of air stream, watt	Mass of air x Air velocity^2 /2	7187
Power of plant, kilowatt	Stream power x Total plant efficiency/1000	3.15

Kinetic energy of air stream is determined by well-known correlation

$$E_{\text{вп}} = \frac{1}{2}MV^2. \tag{1}$$

The mass M of air that travels with velocity V through area S in one second is equal to

$$M = \rho SV, \tag{2}$$

where ρ – density of air (ρ=1,23 kg/m^3) at the temperature t=15C and atmosphere pressure 760 mm of mercury.

Thus kinetic energy of airflow

$$E_{\text{вп}} = \frac{1}{2}\rho SV^3 \tag{3}$$

is proportional to the square of its cross-section and the third degree of its velocity.

1. Modern means of wind energetics.

Wind energy plants, as a rule, transform wind energy into mechanical work of special devices which transform it into different kinds of mechanical, thermal, electrical or other energy.

1.1. Principles of wind flow energy transforming.

A number of physical principles: Drag Principle, Lift Principle, Magnus effect, a principle of automatic vibration and other principles can be used for transformation airflow energy into mechanical work (power).

Drag Principle [9,10] is based on the possibility to get energy from the airflow by moving a body in it. Influence a flow on a body is determined by the projection of pressure gradient along all profile square to flow direction. The corresponding anti-action force which is concentrated in the center of profile gravity is called pressure force of airflow F (Fig 1.1).

If plate is immovable and perpendicular to velocity of the wind the force influences on it

$$F = C_x \frac{\rho S}{2} V_{\text{B}}^2 \qquad (1.1)$$

where: Cx –resistance index depending on the form of a body, ρ – air density - 1,29 kg/m^3, S – square of plate cross-section, m^2, V_B – velocity of airflow, m/sec.

Index Cx depends from the form of a body:

a thin plate perpendicular to the flow Cx = 1.11 - for small plates and Cx = 1.33-1.45 for large plates, for example, square;

a hemisphere, hole facing the flow (parachute) Cx = 1.33;

a hemisphere, the hole is turned in the flow Cx = 0.35;

a body is streamlined drop-shaped Cx = 0.05.

When a plate moves with the velocity *Vn* it runs away from the wind and relative velocity of airflow rushing on a plate reduces. *So* force of pressure will be less

$$F = \frac{\rho S}{2} (V_{\text{B}} - V_{\text{П}})^2 \qquad (1.2)$$

Power equals product of force and velocity of plate

$$N = F\,V_\Pi \qquad (1.3)$$

Power received at the generator equals

$$N = C_x \frac{\rho S}{2}(V_\text{в} - V_\text{п})^2 V_\text{п} = \eta N_\text{в}, \qquad (1.4)$$

where N_Π - power (flow of energy) of airflow, η - Wind Energy Index (WEI). If a plate is immovable in this case useful power equals zero. If a plate moves with the velocity of wind and it is not subjected to pressure and power equals zero too.

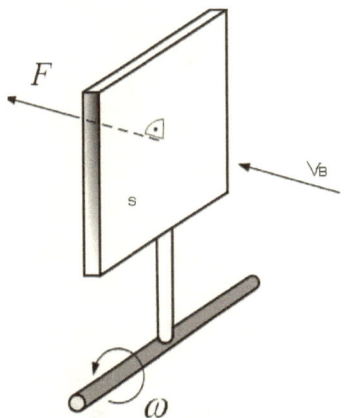

Fig.1.1. Device for using frontal resistance
principle (Drag principle)

For finding the optimal velocity of plate formula for WEI (Wind Energy Index) - the ratio of useful power (energy flow) to total airflow power.

$$\eta = \frac{C_x \frac{\rho S}{2}(V_e - V_n)^2 V_n}{\frac{\rho S V_e^3}{2}} = C_x\left(\frac{V_n}{V_e} - 2\frac{V_n^2}{V_e^2} + \frac{V_n^3}{V_e^3}\right) \qquad (1.5)$$

If

$$\varepsilon = \frac{V_n}{V_\theta},$$

in this case

$$\eta = C_x \left(\varepsilon - 2\varepsilon^2 + \varepsilon^3 \right). \tag{1.6}$$

To find at which meaning ε WEI reaches maximum we need to differentiate η by ε and to equate received a meaning to zero, to make that is to make usual operation to find a maximum. As a result of these transformations we get,

$$\varepsilon = \frac{1}{3},$$

that is for getting maximal η velocity of plate must be three times less than a velocity of the wind. When we put received meaning into (1.6) we get $\eta_{макс} = 0.148\ C_x$. For flat plate $\eta = 0.164 - 0.197$. For semi-sphere with a hole against wind $\eta = 0.197$. Different curve-lined surfaces are used as blades in rotor wind plants.

A principle of Lift Force [9-11] is based on an application of fundamental effect which is used for transformation kinetic energy of air stream into mechanical energy of wind engine. Lift Force is caused by changing of pressures which are formed during a flow of air stream along surfaces. It is shown on the Fig.1.2a as asymmetric profile forms different lengths of streams by directions. Accordingly different velocities of flowing lead to a difference of pressures which creates Lift Force at last.

Profile of wings in the airflow is considered, α - an angle of flow attack. Sum of distributed forces leads to aerodynamic force R with which air acts on moving wing (Fig.1.2b).

Decomposition of force R into vertical Y and horizontal X components produces Lift Force of wing and force of it frontal resistance. As for pressure distribution, one can see on the Fig.1.2a that Lift Force is formed not only because of pressure on the lower generatrix of profile but because of rarefaction on the upper one.

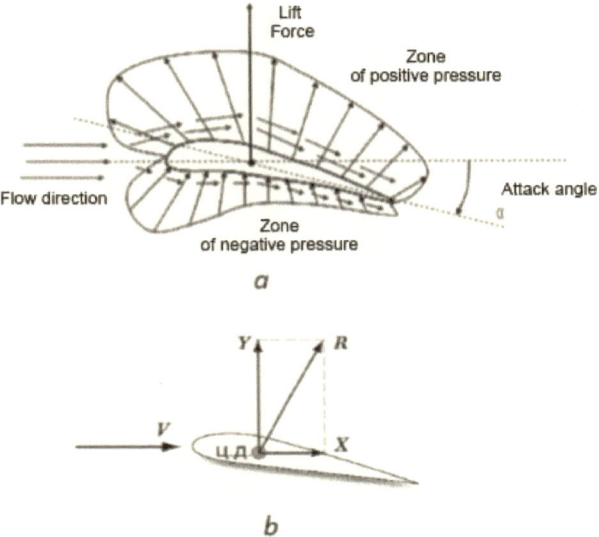

a

b

Fig.1.2. A principle of Lift Force at the example
of aerodynamic profile: *a* - formation of Lift Force,
b – aerodynamic forces of wings.

Profile of wings in the airflow is considered, α - an angle of flow attack. Sum of distributed forces leads to aerodynamic force *R* with which air acts on moving wing (Fig.1.2b).

Decomposition of force *R* into vertical *Y* and horizontal *X* components produces Lift Force of wing and force of it frontal resistance. As for pressure distribution, one can see on the Fig.1.2*a* that Lift Force is formed not only because of pressure on the lower generatrix of profile but because of rarefaction on the upper one.

All depends from these indexes of Lift Force and frontal resistance (see charts of Fig.1.3), which essentially depends on angle of wing attack α.

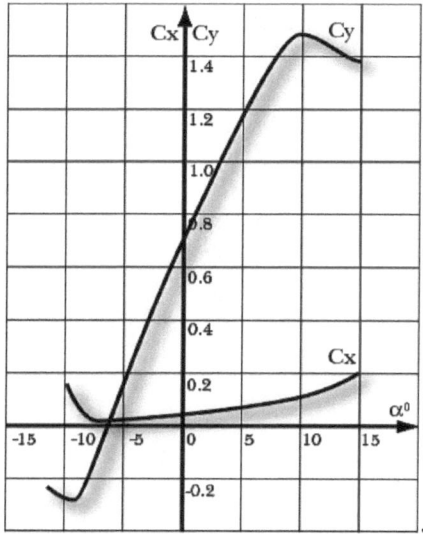

Fig.1.3. Dependence indexes of frontal resistance
and Lift Force from an angle of wing attack.

Magnus effect [12]. According to Fig.1.4 is physical phenomenon arising on flowing around rotating body by a stream of liquid or gas. The force which influences on a body and directed perpendicular to a direction of flow. This is the result of the joint influence of these physical phenomena as an effect of Bernoulli and formation of a frontier layer in an environment around the streamlined object.

Rotating object creates in an environment around itself vertical movement. From one side of the object direction of whirlwind coincides with a direction of streamlining flow and accordingly velocity of environmental movement from this side is increased. From another side of the object direction of the whirlwind is opposite to the direction of stream movement and velocity of environment movement is reduced. Because of this difference of velocities difference of pressures is created. This difference creates transverse force from this side of rotating body on which direction of rotating and flow direction are opposite to this side where these directions coincide. First, this effect was described by German physicist Magnus in 1853.

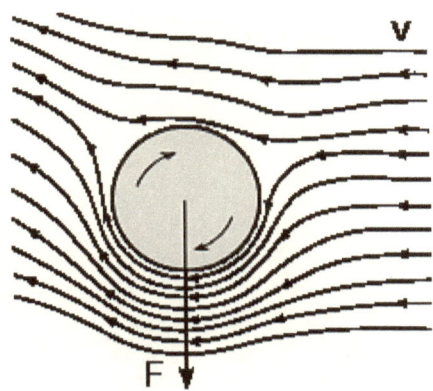

Fig 1.4. Magnus effect during influence
on rotating cylinder

It is known that Lift force of Magnus rotor is perpendicular to stream direction and equals

$$R=C_y(\rho V/2)^2 S, \qquad (1.9)$$

where C_y is an index of Lift Force, ρ - air density, V – flow velocity, S – square diametric section of the cylinder. In particular, C_y=8-10 during linear rotating velocity of cylinder *V(rotation) =(3-4)V*.

A self-oscillation [13]. Repeated movements of system are called self-oscillations when there is source of energy which has no oscillatory properties and from which system receives such amount of energy that is spent. Oscillations in such systems start arbitrarily due to initial fluctuations of oscillatory objects.

Self-oscillations are one of the most distributed nonlinear oscillations of nonconservative systems. They are often used for creation automatic undamped oscillatory systems as for example in watches, piston engines, musical wind and stringed instruments. More often self-oscillations arising in many apparatuses and mechanisms are harmful to normal for work and sometimes even for their integrity. Very dangerous ones are self-oscillations of wings and tail of the aircraft – flutter- which arise at the definite velocities during flight and lead to complete destruction of the aircraft and its wreck. The same applies to building constructions for example to the suspension bridges (Fig.1.5)

Fig.1.5. Self-oscillations of "dancing bridge"
in Volgograd under influence of wind.

On watching the movement of plants, wires, breadth and other objects it is easy to understand that in all these cases undamped oscillations occur due energy of the constant blowing wind. Wherein the oscillatory system produces energy sampling at the right time and in the amount required for compensation of inevitable energy losses. The frequency and amplitude of such steady-state self-oscillations are determined by both parameters of the systems and parameters of its interactions with the wind.

Wind plants in which this effect is used contains elastic (pseudoelastic) working bodies entering into a regime of self-oscillations when they are exposed to airflow.

This oscillatory movement is transformed through special drives in rotation of the working device's axis or otherwise used.

1.2. Wind energy plants with horizontal axis of rotation.

The number of such devices in the world is more than 90%. Several thousand of enterprises serially produce them.

An efficiency of horizontal-axial wind plants is reached only on condition of providing permanent collinearity of the wind wheel axis and direction of airflow [1-5]. The necessity of having a system of orientation to airflow in the structure of wind plant complicates the unit and reduces its reliability. According to experience in the operation of foreign plants up to 13% of the total number of failures fall on the orientation system. It reduces electricity generation and economic efficiency. Using of horizontal-axial plants with propeller turbines is advisable in such cases where the airflow has stable modes – horizontal and permanent from 9 to 18 m/sec.

A tail, as a rule, is used for orientation in small wind energy plants, and electronics with servo motors control orientation in big ones (Fig.1.6).

Exemplary device for a wing wheel is shown in Fig.1.7. On a horizontal shaft wings have been fixed, the number of which in modern wind plants is usually from 2 and more.

The wing of the wind wheel consists of blade on the lever a and blade b fixed on the blade on the lever so that it forms with the plane certain angle ψ. This angle is called the angle of blade curvature. The elements of the wheel are blown by airflow with a relative velocity W at an angle α, which is called the angle of attack, and acts with the force R.

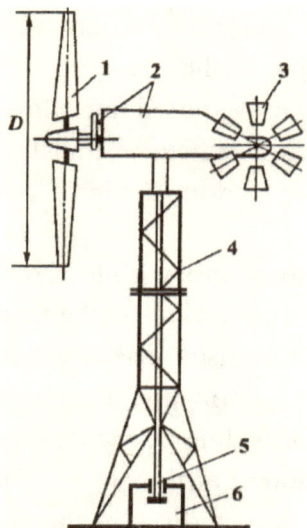

Fig.1.6. The device for controlling the orientation of the horizontal-axial wind plant: 1-wind wheel, 2-transmission-multiplier, 3-windrose, 4-tower, 5 - selection power shaft, 6-generator.

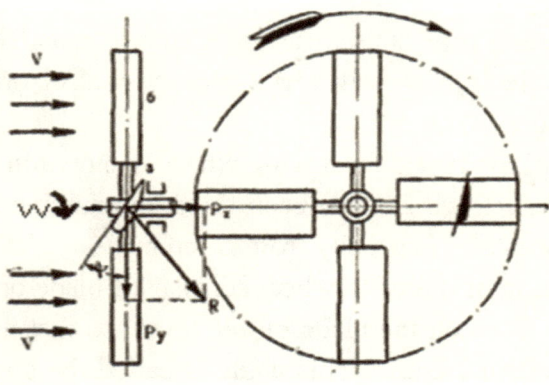

Fig.1.7. Constructive scheme of the wheel.

The angles Ψ and α largely determine an efficiency of the wings. If the force R is decomposed into orthogonal components Px and Py, then Px produces pressure in the direction of the airflow, it is called frontal pressure, and Py acts in the **y-y** rotation plane of the wind wheel and creates a torque.

The maximum value of the force that drives the wheel in a rotation is obtained at a certain value of the attack angle α, that is the angle of inclination of the relative resultant flow ω with velocity W to the blade surface. Due to the fact that the circumferential velocity on the wing length increases with removing its elements from the axis of rotation, relative velocity W of the airflow on the blade is also increased. Together with this, the attack angle α decreases, and when a certain circumferential velocity ωR, where ω is the angular velocity, this angle becomes negative. Consequently, not all wing elements will have a maximum Lift Force. If we decrease the angle Ψ of each blade element than with removing from the axis of rotation so that the optimal angle of attack α was approximately constant, then all the blade elements will work with their maximum Lift Force. Blade with variable angle of blocking has the form of a ruled helical surface. Well done models have an index of air energy using of wing wheel about 46%.

The electric converter [14] of Fig.1.8 is an example of the pilot propeller wind plant. The disadvantage of this design is a small surface of the blade, and as a consequence of this - the initial torque is close to zero, and therefore the launch of such wind engines is difficult.

A velocity of the ends of the blades in strong wind can approach the velocity of sound, creating a noise like a propeller aircraft and interference with electronic devices as well as blades beat birds and other flying creatures.

With a change of wind direction on turning of wind engine arising gyroscopic moment influences on blades and aspires to bend them twice in every turn of blades forward and back, and this can cause large tension which sometimes leads to tearing off blades.

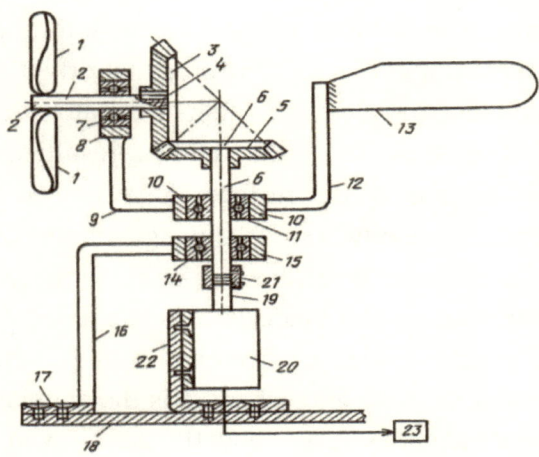

Fig.1.8. Electric converter: 1- blade, 2 - base shaft, 3, 5 - pinion, 4 - key, 6 - drive shaft, 7, 11, 14 - bearings, 8, 9, 15 - bushings, 9, 12 - brackets, 13 - shank, 16, 17- tripod, 18- bed, 19-22 - knot of generator.

Almost all modern wind energy is based on a horizontally-axial three-blade scheme of wind engine. At large sizes, and accordingly large load on the structure, the scheme is dynamic balanced and relatively stable. Numerous studies show that increasing the number of blades more than three does not make sense - WEI (Wind Energy Index) begins to fall due to increase in the resistance of vortex disruptions to numerous ends of the blades. Rotational velocity is reduced.

It requires an installation of more powerful and expensive generator from the point of view as for increasing the rotation velocity of wind generator. To reduce the cost of the generator it is necessary to reduce the number of blades or their area. The latter is difficult to do because of the loss in strength characteristics of the blade. The two-bladed scheme will give an increase in velocity of 20-30% in comparison with three-bladed but reduces its dynamic stability. This drawback completely applies to the three-bladed wind engine. An interesting method for solving this problem - single-blade wind engine scheme (blade with counterweight) [15] on Fig.1.9.

Its velocity is the ratio of the circumferential velocity at the edge of the blade to wind velocity and equals from 10 to 20 units, whereas in the two-bladed wind engines it is 7 to 10, in the three-bladed ones- from 5 to 7.

Fig.1.9. High velocity single-blade wind generator
of the company "Vetrostroy"

However, at low rotation velocity, for example during acceleration there is a dynamic imbalance, so it is difficult to produce reliable single-blade wind generators with a capacity of more than 100 kW.

Horizontal-axial wind generator "Condor Air" [16] - is an example of industrial propeller wind engine (Fig.1.10). The "Condor Air" wind generators have been developed by specialists of the company "Energy Decision", Omsk, just for Russian conditions. Wind plants are produced in the power range of 10-60 kW. Model CONDOR AIR WES 380 / 50-10 is a high-tech wind generator with the horizontal axis of rotation.

Start of the wind plant is carried out at velocity of the wind from 2.5 m/s, and on nominal exit at 9 m/s which allows them to be used in regions with a weak and medium wind. Operating temperature range: -40 Up to +50 degrees in normal performance and from -55 to +50 in performance for low temperatures.

Fig.1.10. Wind generator «Condor Air»

Thanks to using the control unit and three-phase generator, the plants are capable to produce a frequency of 50 Hz constantly over all the revolving rotor range, that is, they can be used without additional equipment - batteries, charging systems, inverters and etc.

1.3. Wind power plants with vertical axis of rotation.

Stages of modern history in the development of wind plants with a vertical axis of rotation are marked by patents on structures that have been successfully used at present [17-19]:

Rotor of Savonius (S.Z. Savonius, Finland, 1922), Fig.1.11*a*;

Rotor Darrieus (J. Zh.- M. Durrius, France, 1931), Fig.1.11b;

Rotor Musgrove (P. Musgrove, Great Britain, 1975), Fig.1.11c;

Rotor "Windsayt" (R. Yutsiniemi, Finland, 1979), Fig.1.11d;

Helical Turbine (A. Gorlov, USA, 2001) which with minor differences is reproduced by turbines «Tvister», «Turby», «Quitrevolution», Fig.1.11e.

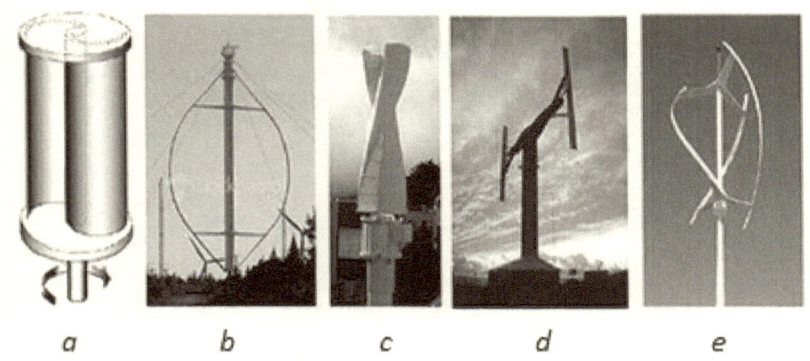

Fig.1.11. Stages of modern history in development
of wind engines with vertical axis of rotation

Vertical axial wind plants were invented later than horizontally-axial ones. Till nowadays it was considered that they are predominantly low-velocity, with the ratio of the maximum circumferential linear velocity of blades to air velocity less than one, which significantly reduces their efficiency, in comparison with horizontally-axial plants, where this ratio reaches 5:1 and more.

The disadvantage of a high-speed vertical-axial plant with the Darrieus Rotor, with rigidly fixed, relatively traverse blades, is a high velocity of airflow at which the rotor self-starting in a rotation, and a lower index of air energy in using [20, 21]. The principle of Darrieus Rotor operation is explained in Fig.1.12.

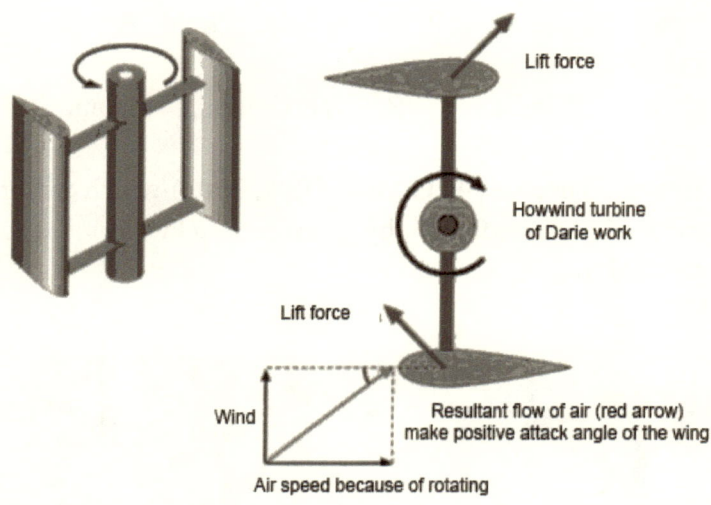

Fig.1.12. Principle of Darrieus Rotor.

Darrieus Rotor is symmetrical construction consisting of two or more aerodynamic wings fixed on radial beams. Each of wings moving relatively to flow is exposed to Lift Force, the value of which depends on the angle between velocity vector of flow and instantaneous velocity of a wing. Darrieus Rotor is characterized by a high coefficient of velocity at low flow rates and at low values of energy efficiency of flow.

Designers are forced to supply such plants with additional devices (electric motor, a rotor of Savonius) to start the rotor and bring it into the working mode. A high velocity of self-launch is caused by a fact that fixed rigid vertical rotor blades in static state cannot create (on flowing around them airflow of medium and low velocity) the required amount of aerodynamic forces on blades and, thus, a sufficient torque on the shaft for bringing the rotor into a rotation.

There is increased interest to these plants in modern wind energetics since the power plants with Darrieus Rotor can reach, according to estimates of experts, 10 - 30 MW. The greatest success in the mastering of Darrieus Rotors with direct blades was achieved by VAWT in the UK (40 kW, 130 kW and more). Wherein Wind Energy Index in such plants is substantially lower than the theoretical limiting value of $\eta = 0.593$.

An advantage of vertical-axial plants is independence of functioning from the direction of flow, possibility of transition from the console bracket main axis to two-bearing, possibility of placing power unit at base of the plant, simplification of the design of blades and reduction of their material consumption, reducing noise and area for its placement, etc.

Plants with Darrieus Rotor can be considered as a competitor of horizontal axis plants. Such rotors have different blade-wing shapes: Φ-, Δ-, Y - and rhomboid.

The question of Darrieus Rotor as a possible alternative to horizontal axial plants has acquired a special relevance when it was discovered that a complex system of orientation devices for horizontal-axial turbines to airflow requires high accuracy and turn synchronization, for not to reduce the efficiency of its operation.

Vertical-axial wind plant "GRC-Vertical» [21] on Fig.1.13, is known as effective design. It has rather a high coefficient of energy flow using - up to 37%.

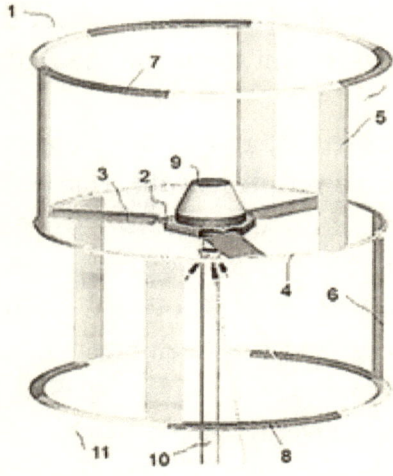

Fig.1.13. Wind turbine "GRC-Vertical": 1- rotor, 2 - hub,
3-fairings, 4- force ring, 5, 6-blades, 7, 8 - rings for blades,
9-casing with a generator, 10-mast, 11-extension.

The advanced Darrieus Rotor is developed in Institute of
Hydromechanics of Ukrainian Science Academy [22] (Fig.1.14).
There are two types of blade control - passive and active. With
passive control, each blade has possibility of turning relative to
the traverse, and if the axis of turn will be near the leading edge,
then aerodynamic forces tend to turn the blade for decreasing
angle of attack α.

This function can also be carried out by special cargoes,
which on exceeding operating velocity and increasing of
centripetal acceleration also turn blades for decreasing of angle α.

If you put on the traverse elastic restraints which turn the
blade, then it becomes possible to adjust passively the blade angle
of attack depending on the velocity of rotor rotation.

With active control of the blades the angles of their rotation relatively to the traverse at each point of the circular path are determined by a special mechanism, which gives ability to rotate the blades of the traverse so that the magnitude and the direction of action of the resultant aerodynamic forces on the blades allow to self-launch wind turbine, even at flow velocities of 3 to 4 m/sec. Moreover, control of a position of the blades enables significantly to improve the performance of such rotor.

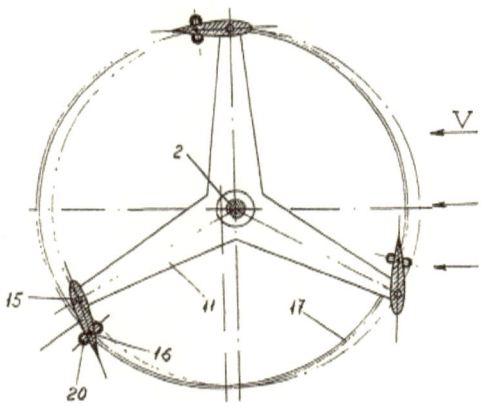

Fig.1.14. Wind power plant: 2-vertical shaft, 11-traverses, 15 - major wing axis, 16-additional wing axis, 17-ring guide, 20-roller, V- airflow.

According to a patent of Ukraine "Wind power plant" [23] the control equipment of this windmill contains the software device that determines optimal displacement value of the center of the annular guide 17, depending on parameters of the rotor. Calculated eccentricity value between the two axes 15, 16 of each blade, which provides the maximum value of wind rotor power with regard to the direction of the airflow. Adjustment of the eccentricity is carried out by two auxiliary electric motors.

Optimization of the work of this windmill is far from strict mathematical justification has intuitive character, but, nevertheless, the efficiency of this rotor is substantially higher than that of the Darrieus Rotor in the classical performance.

Vertical axial wind turbines of "differential frontal resistance" [17-19]. A rigid body of asymmetrical shape (for example, a hemisphere) during different orientations in the flow of air (liquid), with a constant velocity V, it interacts with the flow with different forces, $F_{ЛC1}$ and $F_{ЛC2}$ respectively (see Fig.1.15). The pressure of the wind on the hemisphere oriented to it concave part, more than 4 times exceeds the pressure on the same hemisphere oriented to wind convex part. The cross-sectional area of the bodies is the same. If the hemispheres are fixed on the traverse with two sides symmetrically to the axis of rotation than torque appears during interaction with the moving air mass and the device will rotate with some frequency ω.

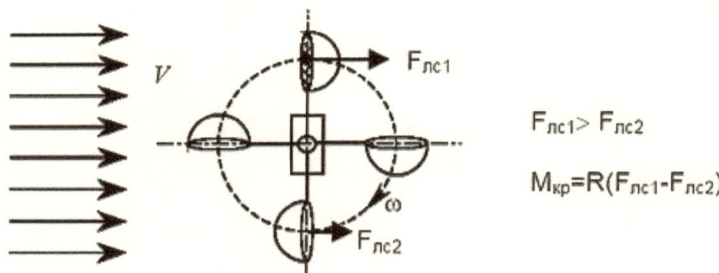

$$F_{лc1} > F_{лc2}$$

$$M_{кр} = R(F_{лc1} - F_{лc2})$$

Fig.1.15. Wind engine operation principle
of differential drag.

The rotor-type wind plants of Fig.1.16 have the lowest efficiency of all previously mentioned. Their work is also based on the difference in pressures acting on the concave and convex surfaces of blades.

The advantage of this plant is its ability to work at low air velocity, a simplicity of its kinematic scheme due to the lack of necessity in an orientation of the blades, and as a consequence low cost. Another feature of this scheme is the wider swing of the blades, the more power we can receive at a constant velocity of airflow. Due to this plants of this type are capable to work at flow rates from 0.5 m/s.

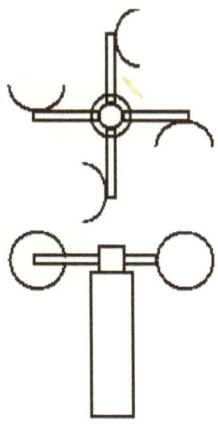

Fig.1.16. Rotary type wind plant

This is achieved by the fact that the arms of the blades act as a lever. Thus the longer the bracket the greater the lever and less effort we need to turn a shaft.

Savonius Rotor [24,25] is made in the form of half cylinders, but they are located somewhat differently, then in rotary windmills. As can be seen from Fig.1.17 airflow is used more rationally.

The torque appears when the rotor is flown by airflow due to different resistance of the convex and concave parts of the rotor. The wheel is simple but has a very low WEI (Wind Energy Index) - just 0.10 - 0.15. A number of original designs are developed on the principle of frontal resistance also.

Carousel wind wheel. Wind-wheel [26] (Fig.1.18) contains L-shaped elements mounted on the vertical axis and consisting of a radial blade on the lever and the similar blade on the lever in construction additionally established perpendicular to the radial one fixed in blades on the lever with the ability to rotate thrust frames.

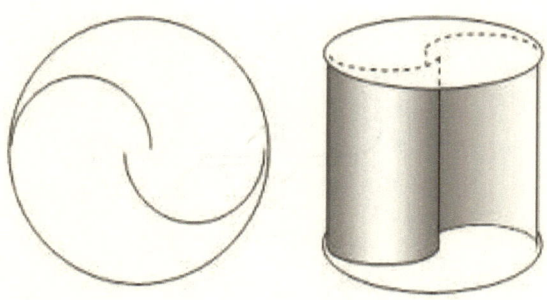

Fig.1.17. Savonius Rotor

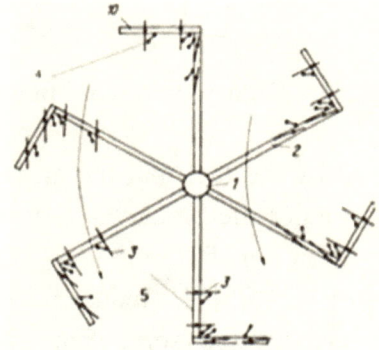

Fig.1.18. Carousel wind wheel: 1- vertical, main axis,
2 - the radial blade on the lever, 3 - blades,
4, 5 - stops, 10-panel.

Wind-wheel works in 3 quadrants relatively airflow vector, the blades on the radial panels work in the 1-st and 2-nd quadrants, and blades on the perpendicular panel - in the 2-nd and 3-rd quadrants.

Disadvantages of this wind wheel are unevenness of rotation because sets of L-shaped swings are parallel, a complexity of the construction consisting of a large number of nodes and parts, low reliability during operation and negative ecological indicators (destruction of birds and other flying creatures).

Wind-driven rotor-slotted type [27], Fig.1.19 is a tube with two slots insides, which rotates the rotor. Example of such plant can be the turbine on the ship "Alcyone" of famous researcher Jacques Yves Cousteau. Advantages of plants of this type are their resistance to a large velocity of incoming airflow, due to their streamlining, and as a consequence, low metal consumption, and, consequently, cost.

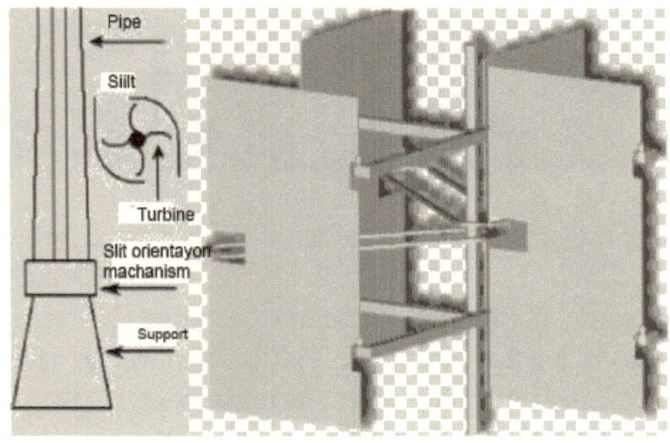

Fig.1.19. Rotor-slotted wind plant.

The disadvantage of such plant is low efficiency due to the fact that the incoming airflow falls in a vertical rather narrow slit. The problem of slit orientation to the airflow remains the same as for propeller wind plants.

Combined wind plants with vane wheel on a vertical shaft - perspective designs which become more widespread. Although they lose a little bit in an index of air energy using in comparing with classic horizontal axial structures due to energy losses when turning the airflow, at the same time acquire the advantages of above mentioned vertical-axial plants.

Wind turbine IMPLUX [28]. Australian company Katru Eco-Energy has developed a new type of wind turbine (Fig.1.20), designed to capture airflows on the roofs of high-rise buildings.

Fig.1.20. External view of the IMPLUX wind turbine.

A rotor of the "omnidirectional" turbine is located on the vertical axis on the top of the IMPLUX housing. Horizontally oriented blades of it are given in movement by the ascending airflow (Fig.1.21).

The central chamber of the turbine with profiled blades instead of walls that let in airflow in central camera but thanks to the penetration angle release it from the back side but cause it to flow upward inside the camera. Thus, at least 87% of airflow reaches the top of the hull with located their horizontal blades of rotor which shaft's rotation produces electricity.

The result is a turbine with a single moving part - rotor. This reduces operational costs and also does the work of

IMPLUX quieter than conventional horizontal turbines with rotors of the same dimensions.

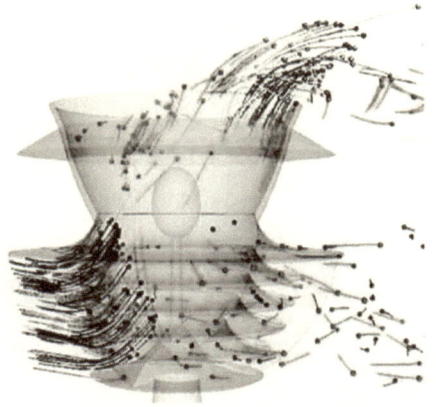

Fig.1.21. Airflow scheme in the turbine IMPLUX

Unlike other similar devices, the turbine IMPLUX can use any wind orientation, without changing its own position and/or orientation.

AeroGreen Wind plant of the new generation developed in a joint project of Irkutsk State Technical University and joint-stock company "Aeroenergotech" (Fig.1.22) [29], uses a vertically-axial aerodynamic turbine instead of traditional three-bladed scheme. The high performance of the new generator was achieved through the replacement of traditional blades on short blades of polymer material. As a result, the wind plant does not need to be directed to the wind that gives great advantage and significantly reduces the cost of energy.

Another plus of the new generator was his protective noise-absorbing housing. Without large blades, the generator produces so little noise but in special housing work of the wind wheel becomes almost noiseless and safe - this allows to install the generator anywhere and not collect preliminary numerous concurrence of neighbors that Is especially important for Europe.

The operation of the plant is independent of wind direction and adverse weather conditions (hail, rain with snow, hurricane). The structure is made of light polymer materials, the basis of which may be plastic packaging, packaging, bottles, plastic bags, etc.

Fig.1.22. The wind turbine
of the new generation "AeroGreen".

This plant can be used as an alternative source of electricity on various transport means, as an onboard power station, in private houses, on enterprises and gas stations for recharging electric vehicles. By AeroGreen plant is commensurate with the cost of a traditional wind plant, but it produces twice as much electricity.

The essential feature of the above structures is the presence at the outlet of the turbine special fairings, which create a vacuum in horizontal flow of air above the turbine, due to an effect of the wing and accordingly accelerate the flow of air through the turbine.

1.4. Wind power plants without blades.

The bladeless wind power plants unite a wide class of devices the action of which is based on a variety of principles, effects, and methods.

The Magnus compound Rotor [30] of Fig.1.23. As a technical result, we have an increase of Lift Force by 30-50%, enlarging of direct control of the Lift Force and decreasing of energy consumption for rotor rotation.

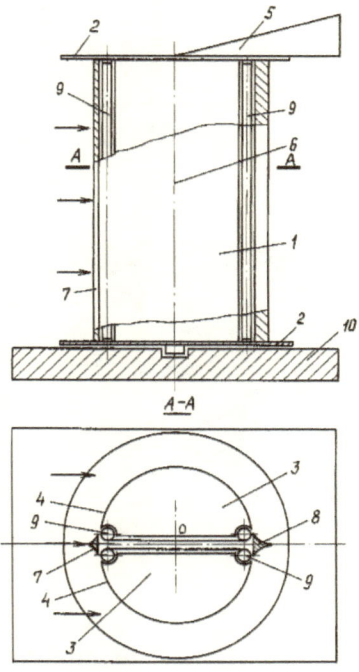

Fig.1.23. The design of a windmill based
on the effect Magnus. 1-rotor Magnus, 2- end washers,
3-cylindrical body, 4 - side shells, 5-weather vane, 6-central
longitudinal axis, 7-flow divider, 8 - flow divider, 9-rollers,
10-base.

This result is provided by the fact that in a compound rotor containing a cylinder and end washers, the cylinder is made of several cylindrical bodies with a height which is equal to the height of the cylinder. Each of the bodies has a lateral closed shell with possibility of rotation, the body is fixed to the end washers, rotor is equipped with a weather vane and can be rotated around its central longitudinal axis, and between cylindrical bodies along their entire length in the front part of rotor flow dividers are placed and in the rear part flow diverters. The rotor is used in the fleet as an auxiliary propulsion device.

The wind generator-sail [31] on Fig.1.24. Another one option without blades - wind plant created by a principle of the sail. Another one option without blades - wind turbine, created according to a principle of the sail.

Fig.1.24. The wind generator-sail

Outwardly it looks like an ordinary wind generator with a horizontal axis of rotation, only instead of blades, it has the device resembling a satellite dish. This is the sail which catches wind. The wind makes the sail to oscillate, the oscillations drive the pistons and switch on a hydraulic system, and thus kinetic energy is converted into electrical energy.

The developers claim that such wind generator is more reliable, durable, noiseless and productive and it does not require special expenses for maintenance - spending is reduced by 45% compared to conventional wind plant.

Wind plant Vortex Bladeless [32] on Fig.1.25. This new wind turbine swings in the wind, creating electricity without rotating blades.

Principle of extracting wind energy by means of cylinder or cone driven by the wind is used

Fig.1.25. Vortex Bladeless Wind Plant

It is known that inflow passing near axisymmetric object (for example, a cylinder or a cone) vortex path is formed behind it - double vortices that are formed and come off not simultaneously. As a result direction alternating force will act in a transverse direction to the flow.

This force can if the object is fixed on the movable fastener oscillate it. If we correctly organize the system of mechanical energy removing then this mechanical energy can be transformed into electrical one. Engineers Vortex Bladeless have developed a generator that uses this mechanism. Thin cone -shaped turbines are made of carbon fiber and fiberglass with a motor on the bottom, and not from above, like traditional turbines for improving strength. Ring magnets on the basis of a cone give a momentum of rotation regardless of wind velocity. Vortexes of winds rotate synchronously along the entire length of the cone.

Energy Stems [33]. Design company Atelier DNA proposed the idea of generating energy with using stems made from impregnated resin carbon fiber. A diameter of wind stems at the base is 0.3 m. To the top, the design narrows to 0.05 m. Each wind stem contains electrode layers and ceramic discs made of piezoelectric material, with compression generating current. Compression is a consequence of self-oscillation of the stem in the wind. These wind stems are very effective because they do not have friction losses in comparing with mechanical systems like windmills.

Vortex wind power plant [34] (Fig.1.26) serves to convert the airflow energy into the electrical energy.

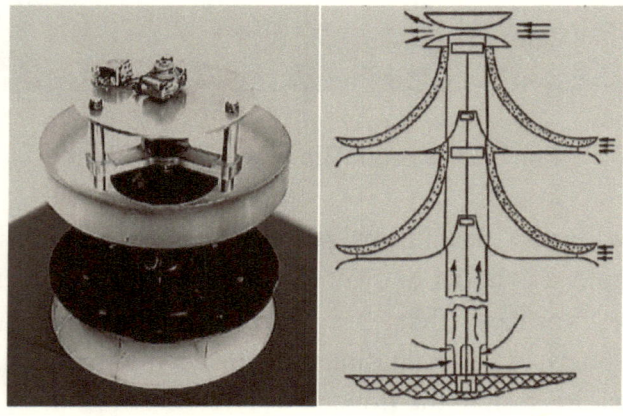

Fig.1.26. Vortex wind power plant.

Know-how of plant - "generator" Vortex is a device that converts uniform flow into Vortex jets, and being a concentrator power, organizes and accumulates energy of airflow similar to the way in natural conditions when kinetic energy of the airflow distributed in a significant amount of flow concentrates to significant values in a compact core of the natural tornado

According to the calculated data operation of a plant is effective in the range of operating flow rates - 3 – 40 m/sec. Calculation characteristics of the vortex plant for power 5.0 kW: diameter – 3.0 m, height – 1.8 m, weight - 120 kg. Estimated annual electricity production – 32620 KWh.

Compared with traditional lobed plants, a vortex wind plant according to Fig.1.26 provides: possibility of operation at lower (in 1.5-2.0 times) working wind velocities, reduction of weight and size characteristics, simplicity and reliability of construction (there is no shaft and the impeller, the system of orientation to the wind, the number of rotor rotation is controlled by changing the area of air intake), high energy efficiency of airflow (up to 0.40).

Wind generator which works by means of water droplets [31] according to Fig.1.27.

Fig.1.27. A drip wind generator.

In addition to standard vertical and horizontal wind plants already mastered by serial producers, there are now other types of wind generators. For example, Holland wind generator which does not have moving parts. It is considered that it will be distinguished by increased reliability.

Tubes with special nozzles and electrodes are placed in parallel inside the metal frame. Positively charged drops of water move from nozzles. These drops are blown to positively charged electrodes by the wind as a result of this the charge increases

Productivity of the device depends on the number of drops, wind velocity, and electric field strength. Such wind plant differs from the standard models by an absence of moving and rubbing parts, hence - no noise, vibration and wear due to friction.

1.5. Scaling of wind power stations.

Wind energetics takes the second place in the world is among alternative sources of energy. The main problem restricting its wider use is relatively high cost as compared with traditional sources of energy (hydro, heat and nuclear power plants). One way to overcome this drawback is increasing the power of individual plants. The power developed on the axis of propeller wind plants is proportional to the square of its diameter and cube of air velocity and depends on its size [9-11] (see Fig.1.28). Limit of a power of classical horizontal-axial structures - up to 10 MW.

Large plants often have to be untwisted from an extraneous source. It should be noted that the velocity at the end of the blades with a strong airflow approaches velocity of sound, creating a noise like a propeller aircraft and interference with electronic devices and also blades beat birds and other flying living creatures.

When turning wind turbines with changing a direction of airflow significant gyroscopic moment acts on blades. For compensation the negative gyroscopic effect the blades are lightened maximally and special devices are used – windroses carrying out a very slow turn of the wind engine. Windroses system and other additional devices greatly complicate the design of the wind engine.

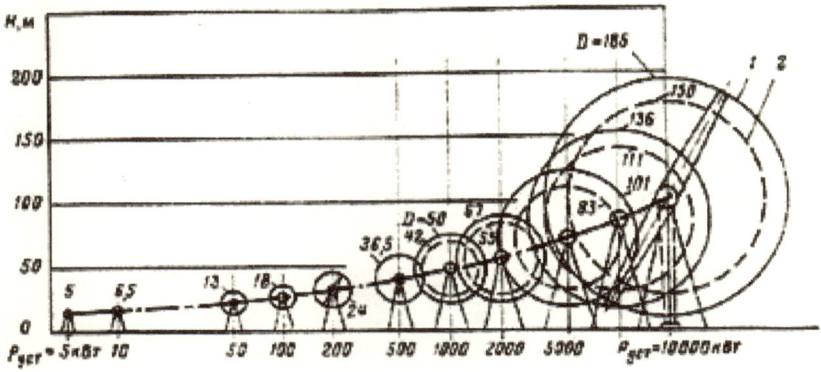

Fig.1.28. Wind power plants of different sizes
at flow air velocity 17.6 m/sec.

The blades of the winged multi-bladed (18-24 blades) slow-moving wind turbines are made from thin, slightly curved sheet and they are mounted on the wind wheel with a horizontal axis of rotation. The merit of this construction - a great initial moment, launch without help. The plant starts at the wind velocity of 2-3 m/s, it does not make noise. Disadvantages: Efficiency is lower in compared with small-bladed wind turbines; Metal consumption is increased, there is a gyroscopic moment. At-large windrow wheels, windroses are also used.

Germany takes the world's leading place in the use of wind energy. Developers of Germany, seeking to increase the capacity of wind turbines went along the way connected with increasing size of traditional propeller plants. Their latest models are equipped with three-bladed rotors which weight many tones with a diameter of up to 66 m. and capacity up to 2 MW.

Enercon E-126 - wind generator produced by German company Enercon [35] on Fig.1.29. The capacity of the power plant is 7.58 megawatts. A height of tower (from the base to the axis of the rotor) can vary in depending on requirements, in the standard version it is 135 m, the span of the blades is 126 m, the total height of plant - 198 m.

Depending on the weather conditions, the power can exceed the nominal one, anticipated annual production in Enercon Electricity ≈18 million kWh.

Fig.1.29. Wind generator Enercon E-126

The weight of the foundation of the installation should be 2500 tons, the weight of the supporting tower is 2800 tons, the gondola generator has weight 128 tons, weight of the generator - 220 tons, a weight of the rotor together with blades - 364 tons. Total weight of the wind farm is about 6000 tons.

Model of the wind power plant [36] (Fig.1.30) allows to increase the efficiency of wind generators and reduce the cost of electricity from two till three times. It developed in Far Eastern

Federal University. Such plants are proposed to place on the water surface.

The wind power plant represents a large-sized construction in the center of which there is a low tower above water and rotor with blades rotates slowly around. Energy is transmitted through traction on the central hub connected to the generator. The wind plant is kept on the surface of the water on the pontoon.

Horizontal stability of the structure is ensured by anchors on the seabed. Rated diameter of a typical plant with a capacity of 10 MW is 200 meters, the height of the blades- up to 40 meters.

Fig.1.30. Marine Vertical-Axis Wind plant

One such plant can provide with electrical energy up to five thousand homes. Such plants are also suitable for the supply of remote coastal settlements, for example in the Magadan Region, the Kuril Islands, Islands of Sakhalin or Kamchatka. Proposed technology if it is necessary allow, to tow such plant on water.

Not only the Far Eastern scientists are moving to the direction as for using of marine turbines with a vertical axis that confirms that they work in the right trend. Fig.1.31 shows the project of American company Offshore Energy [37], where it is considered that the new wind turbines can have much more power than classical ones

Fig.1.31. Floating Sailing Wind plant

It's like a ring of yachts, where the blades are sails. You can create a wind generator with a capacity of up to 100 MW since the technological constraints for this do not exist.

Vertical axial wind power station «Aerogenerator». British Architectural Studio Grimshaw Architects together with Windpower Ltd has developed an original wind power plant "Aerogenerator" [38] (Fig.1.32).

It according to the figurative expression of the authors represents the hybrid of glider-pencil and harp in the form of the letter V.

The height of the wind turbine with the vertical axis is about 140 meters. According to the idea of the developers, the unit can be set in the sea at a distance from the shore, here strong winds blow. By its design, the "Aerogenerator" is a modification of the widely distributed Darrieus Rotors, used in wind plants of significantly smaller sizes.

At velocity of three revolutions per minute, one Aerogenerator plant can produce 9 megawatts electricity (for comparison: conventional power industrial wind plants - 2 megawatts).

Fig.1.32. Vertical-axial wind power station
«Aerogenerator»

No less important - aesthetically new units look much more interesting than previous ones. Waine Billings, one of the developers of the architectural component of the plant, speaks about the wind plant as follows: "We saw it as a figurative element that might appear in the entrance to the harbor or to the industrial area. Moreover, the turbine should not be invisible at all."

Multi-modular wind power plants consisting of dozens of small wind turbines [39] according to Fig.1.33 are one of the main directions of development wind energetics.

The reason for the most shortcomings of horizontal-axial wind plants, leading to loss of power, consists in the special nature of the flow around their ends of blades by air stream.

To change it the developers set the wind wheel (propeller) of the module into the ring fairing and fix it on the central body with the help of profiled blades (Fig.1.34).

Fig.1.33. Towers consisting of wind power modules

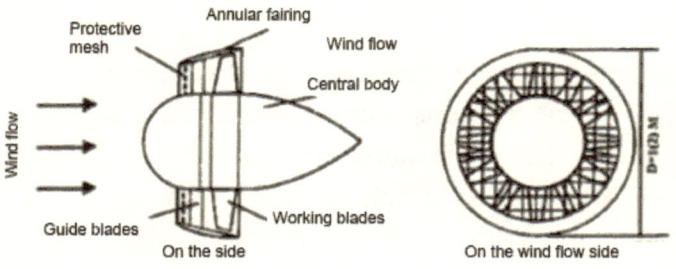

Fig.1.34. Wind engine-module
with an improved aerodynamic scheme.

The blades (or guiding device) are designed so that a preliminary twist of the wind flow is created on the way to the wind wheel. Thus, twist at the exit from it and related phenomena of power loss reaching in conventional wind plants 5-10% are eliminated. The ring fairing significantly reduces the noise of working wind wheel. At the entrance it is closed by a grid, providing protection of birds.

Despite the fact that the power losses in comparing with the settings of the traditional scheme increase for friction account from 7 to 20% (in proportion to the increased area of streamlined surfaces), total losses of power are reduced by about half. This succeeds to achieve through the application of more aerodynamic scheme.

Structurally, the module is more complex than traditional plants but has more energy efficient. Test results in aerodynamic pipe show that it provides a higher coefficient of wind energy use than other modern wind plants. Another advantage of a module is the greater efficiency of power take-off at the moment of launch due to which it needs much smaller starting velocity. Modular wind energy plant is not only cheaper, more environmentally friendly and more economical than propeller ones but also more convenient to operate.

Powerful wind power stations represent a further development of wind plants using. First of all wind power is an opportunity of the power supply through affordable, extremely capacious and environmentally safe resource - airflow.

Industrial wind plants differ in size and capacity, as well as adaptability to the terrain in which they will be placed: hills, plain, coastal zone, sea bays. The most important factor is the average velocity of airflow in the place where wind plant is installed.

It is natural that the biggest wind potential is observed on the sea coasts, on the Uplands and mountains. Nevertheless, there are still many other areas with the potential of airflow, sufficient for its use in wind energetics. Wind stations require open spaces with frequent and uniform airflows.

Wind stations can include from several units up to several hundred located side by side power plants (Fig.1.35). These stations have a particular advantage in conditions of new construction on undeveloped territories. They are able to solve the problems of energy supply, both large settlements, and small settlements.

Fig.1.35. Stations - wind parks

Wind plants receive energy of airflow and at the same time reduce the flow velocity behind the wind turbine, and also form swirls of the flow. If plants work In one direction of the airflow, the second turbine will receive a reduced flow rate and will not work optimally, because of the turbulence caused by the first turbine. Caused by this the expansion of airflows called Wake-effect, and has a significant impact on neighboring plants that influence on the production of electrical energy [40].

Fig.1.36*a* shows the weakening of the wind flow behind the first turbine. The corresponding curve in Fig.1.36*b* shows the difference in the output power of the first and second turbines.

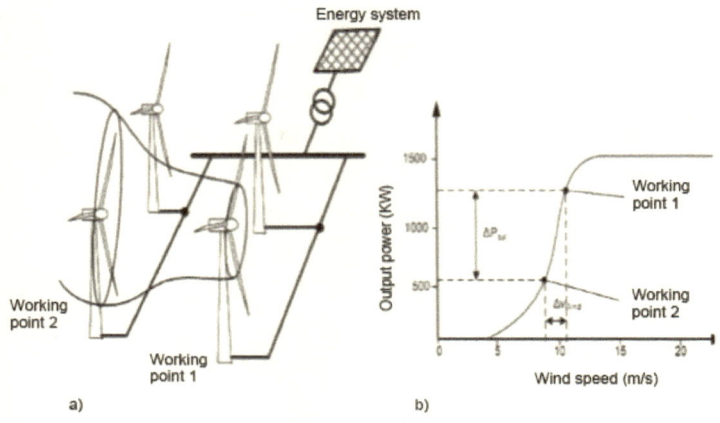

Fig.1.36. The Wake-effect principle (*a*)
and the curve relative power of the wind turbine (b).

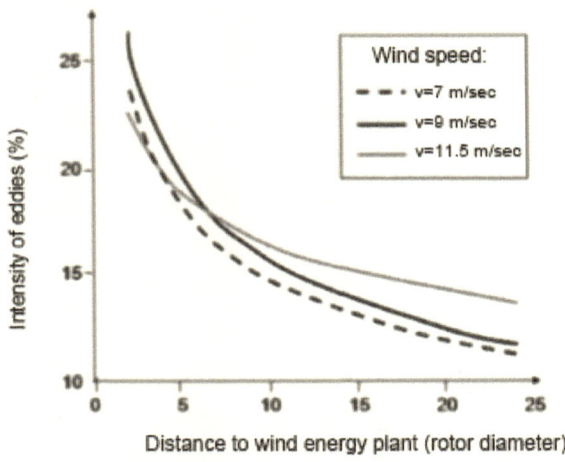

Fig.1.37. Intensity of turbulence depending on distance

The intensity of turbulence can be described   relative to the distance between neighboring   plants and in addition, depends on the velocity of the airflow according to Fig.1.37.

Thus, the space between the wind plants should be large enough so as an influence of turbulence was insignificant. Also, it is additionally increased the technological service life and energy output.

The option of wind plant taking into account the effects of wake and turbulence is given in Fig.1.38.

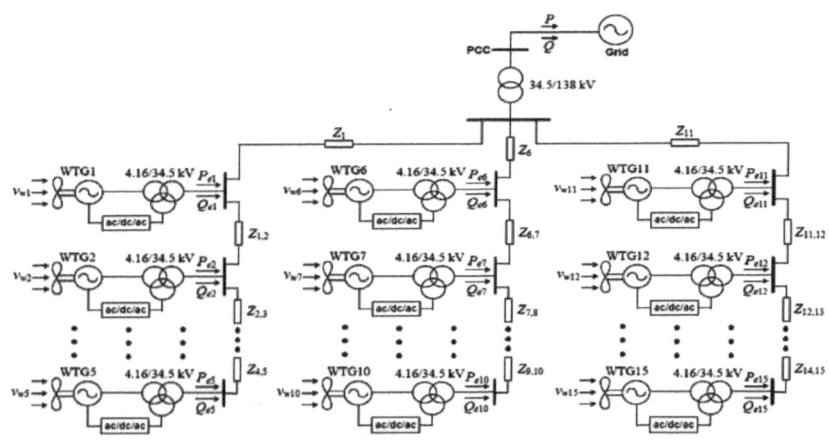

Fig.1.38. The configuration of a wind station
equipped with 15 DFIG wind turbines
connected to the electrical network.

An important quality of the wind farm is that plants can be connected to a common electrical network and transmit the generated energy to it, or can be autonomous when the consumer is close to the plant.

The wind station – "A forest of wind stalks" (Fig.1.39). Design Bureau Atelier DNA [33] proposed the idea of "Green" power supply free from cars. The city Masdar, not far from Abu

Dhabi in the area of 6 square kilometers. For the city's energy supply it is proposed creation of a "forest" of swaying in the wind energy stalks generating electric power due to piezoelectric effect.

Fig.1.39. Energy-efficient stems of Atelier DNA

Atelier DNA believes that there will be enough 1203 wind stalks for Masdar, separated by 10-20 meters from each other, each of 55 meters high, with a concrete base.

1.6. Wind energy regulation systems.

In wind power plants with pitch regulation (with attack angle regulation) [41], blades can actively be rotated around its axis changing the output power. The disadvantage is more complex construction, because of the mechanism for pitch and the management controller. The pitch control is shown in Fig.1.40.

Since the blades of pitch-regulated wind plants are turned into the wind to reduce the output power, blades with active control of velocity loss are rotated in the opposite direction.

Variable Velocity Wind plants [40-43] became the most common among the installed units. They allow you to achieve the optimal output power over a wide range velocity of airflow by adjusting the rotation velocity during changing of input wind velocity.

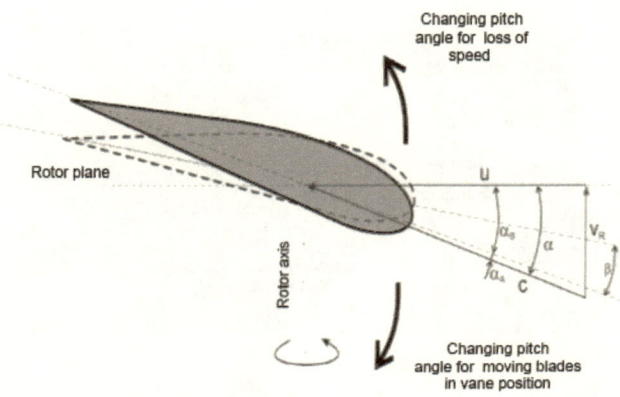

Fig.1.40. The principle of active power control - pitch control.

Typical for this group is that the generators are not connected directly to the network, but with the help of frequency converters. This helps to reduce coupling of the network frequency, and frequency of the generator and, consequently, to regulate the velocity.

Nowadays   two main concepts for wind plants with variable velocity of rotation: the first one is that velocity control with the help of asynchronous generator with dual power supply (doubly fed induction generator - DFIG, Fig.1.41), and the second one - velocity control with synchronous generator with direct rotation (direct driven synchronous generator - SG, Fig.1.42). Both of these concepts use pitch control as the main aerodynamic concept to limit the output power.

In a single-line scheme based on asynchronous doubly fed induction generator asynchronous generator with a current collector, rings are used. A three-phase winding of the stator of the generator is connected directly or uses a transformer with access to the network, while rotor connected to the network by means of the electronic power converter.

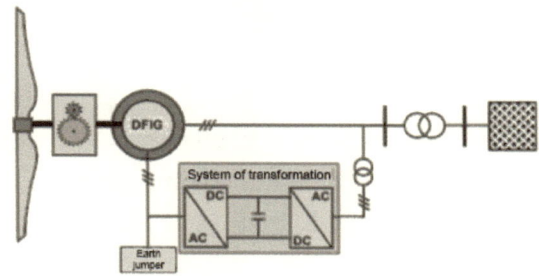

Fig.1.41. General diagram of the wind turbine
with asynchronous doubly fed induction generator DFIG.

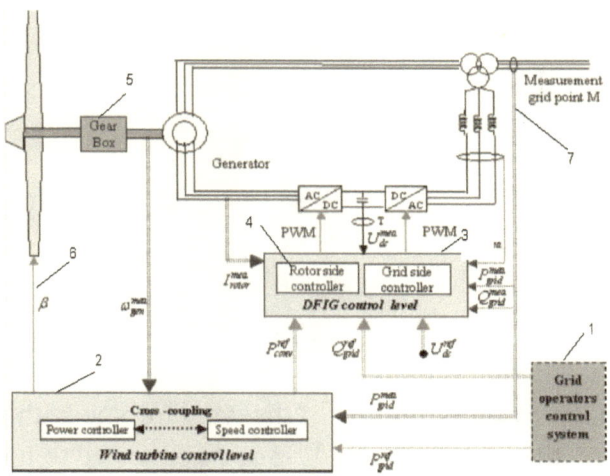

Fig. 1.42. DFIG system configuration: 1 - node of operational monitoring of this system, 2 - node of level control of power-velocity, 3-controller from the network side, 4-controller with side rotor, 5-reducer, 6 - control of angles of rotor blades, 7- parameters of the mains voltage.

The converter device allows you to change angular velocity in a certain interval. This range is limited both by the mechanical parameters of the wind plants and technical characteristics of the converter device. Most plants with DFIG contain reducer which allows connecting the slow-moving shaft of the plant with a high-velocity shaft of the generator.

In some cases to stabilize wind power plant, when the network unevenly consumes energy, energy storage, in particular, supercapacitors, are used. The main advantage of electrochemical condensers in comparison with analogs is their potentially higher electrical capacitance at comparable power. For modern electrodes based on oxides and hydroxides, the capacity of 600-700 F/g is not the limit. Energy storage system supercapacitors (ESS) are controlled by controllers and is coordinated by the supervisory network of the wind park, while the lack of active power is compensated by the ECC block (Fig.1.43).

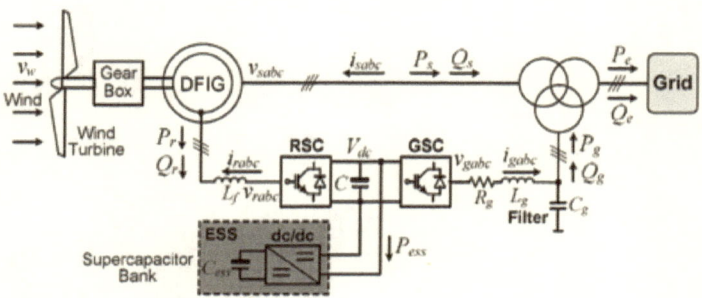

Fig.1.43. The configuration of the wind turbine with DFIG equipped with supercapacitors ESS and connected to the mains.

Fig.1.44 shows a general diagram of concept using a synchronous generator. A stator of the generator is connected to the network using converters. This allows the turbine to function with a regulated angular velocity of the generator. If the angular velocity of turbine changes from the nominal value due to the changes in air velocity, a frequency of induced voltage on the generator of wind plant is not constant and does not equal 50 Hz. Therefore voltage of generator must first be transformed with the rectifier in the DC voltage and then with help of inverter into an AC voltage with constant network frequency.

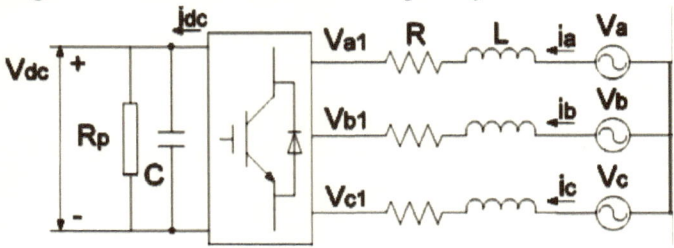

Fig.1.44. Connection diagram of the converter to the network

Frequency converters for wind plants with synchronous generators must mat the rated power of the generator. Thanks to the complete elimination of the generator and network dependence in wind plant, its electrical parameters at the point of general connection, such as voltage, frequency, can be better adjusted.

Fig.1.45, 1.46 shows a general diagram of wind plant constant in velocity with an asynchronous generator with a short-circuited winding of the rotor.

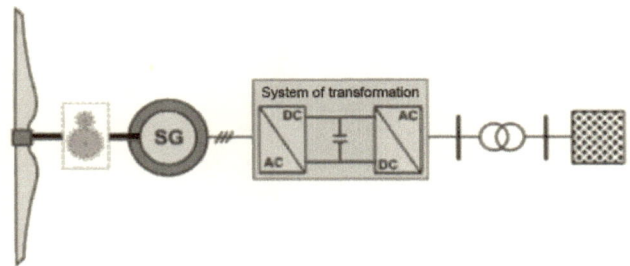

Fig. 1.45. General diagram of the plant
with synchronous generator of direct transmission rotation

With the complete elimination of generator dependence, the corresponding plant's operation becomes more flexible and electrical parameters at the point of common connection, such as voltage, frequency, can be better adjusted.

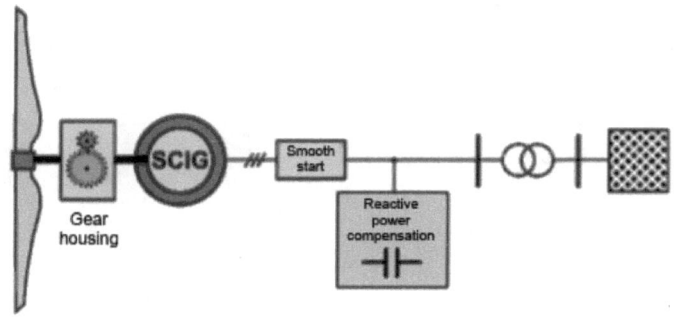

Fig. 1.46. General diagram of the plant with
an asynchronous short-circuited winding generator.

If the wind velocity becomes too high and exceeds the rated value then the output power of turbine and its angular velocity should be minimized in order not to exceed nominal value and do not damage the wind plant. Reduction of output power can be achieved by increasing the pitch angle which is set by the controller.

Summary of the first chapter.

The purpose of modern wind power systems is, mainly, the transformation of the kinetic of wind energy (airflow) into electricity. Devices of wind power generation used for this purpose are very diverse, however, all of them, with rare exceptions, are built on the principle of mechanical extraction of wind energy with the subsequent transference of mechanical energy to the entrance of generator.

More than 90% of the world wind power park include propeller plants with a horizontal coaxial configuration of the windmill and generator. They are perfect high-velocity devices with a high degree (coefficient) of the airflow energy extraction. However, their perfection has a downside and more likely connected with forced measures for compensation their shortcomings.

The main structural elements, in the first turn working blades of the wind engine due to the layout and aerodynamic features are cantilever structures of small cross-section, which leads to dynamic imbalances and instability, as in separate nodes and in the wind plant as a whole. This leads to decrease in reliability and durability of plants. The most vulnerable element of such units is the turn node of the power unit's axis to a direction of wind because of significant gyroscopic moments. Overcoming these shortcomings are achieved through the use of new more complex structures and materials that leads to a significant rise in the cost of plants.

One of the directions as for development of horizontally-axial structures is the transition from propeller wind engines to compact high-velocity turbines with aerodynamic concentrators of airflow.

Following the horizontal-axial wind turbines with big advantage in terms of application, there are plants with vertical layout of main axes of power unit. They appeared much later, very diverse, their experience of study and practical application is significantly inferior as compared with horizontal axial plants, however, according to experts, they have good prospects in wind energetics.

Their main advantages are related to the features of vertically-axial configuration. First of all, this is better dynamic balance and stability due to the lack of turn devices to the direction of the wind and, accordingly, gyroscopic effects. As a consequence - greater reliability and durability, the possibility of using more simple and cheap construction, components and materials The location of the energy-power nodes makes it possible better access and simplifies maintenance.

The simplest constructions of vertical-axial wind engines are based on the principle of differential frontal resistance. They are easy to manufacture and they are cheap, but they have very low extracting flow energy index and therefore are used in inaccessible places or as auxiliary devices in large wind plants.

As the perspective direction of vertical axial plant development, it is considered using of wind engines such as Darrieus Rotor. It's good balanced high-velocity rotor with fixed vane blades, relatively easy to manufacture and maintain, but indicators of its energy efficiency have contradictory character and need serious analysis. Finally, vertically-axial rotors with adjustable blades. It is considered that these wind engines have relatively high factors of wind energy extraction, but not able to work in high-velocity modes. Open information on computational models and techniques for such devices is severely limited, so they need to be studied.

It makes sense to note a special category of vertically- axial plants in which the traditional horizontal-axial wind engines are used in vertical linking. This requires vertical redirection of wind flow, however, using of modern turbines with flow concentrators allows you to limit losses of pressure and ensure high energy efficiency.

When comparing different wind plants, bear in mind that propeller plants are capable to be started and work effectively only with strong winds of 13-15 m/s. That considerably narrows territorial and time limits of their using. In addition during operating mode, they create strong electronic interference and high noise background. In this respect vertical-axial plants are more universal - they are capable to work with winds in the wide range of intensity and ensure environmental safety.

Possibilities of power plants scaling are the main factor in increasing their effectiveness. Most expert assessments are reduced to the fact that possibility of increasing the power of an individual horizontal -axis wind power plant due to the growth of its overall dimensions are exhausted at the level of 10 MW. For vertically-axial marine plants, there are no principal obstacles for bringing this figure even to 100 MW.

Means of increasing efficiency of wind power plants along with optimization of structures and scaling is using of optimal regulation systems of the regime parameters. Regulation is carried out at all stages of energy extraction and transference: in the wind turbine by means of aerodynamic shape optimization and orientation of the working blades; in generator by means of load variation and energy accumulation and in the transmission of electricity to the network at the expense of overflows and other means of network regulation.

2. The theoretical basis of wind turbine optimization.

Energy potential of wind (airflow) is determined by the so-called specific power - kinetic energy of the flow passing at time unit through its unitary orthogonal cross section. This indicator depends solely on the flow rate and is proportional to the cube of velocity (see Table).

Wind velocity, m/s	2	4	6	8	10	12	14
Specific Power, W/m	4,8	38,4	129,6	307,2	600,0	1036,8	1646,4

This potential determines the natural boundaries when wind energy are extracted by wind engines. An optimization problem is determining the conditions of wind engines functioning – values of constructive and regime parameters that provide a maximum possible approximation to the limiting airflow potential.

When studying the optimization possibilities one should proceed from the fact that almost the entire segment of industrial and small-scale power plants using wind energy is occupied nowadays by the so-called horizontal-axial and vertical-axial wind turbines. On the other hand, there are plants in which wind flow is concentrated and turned in the vertical direction. In such wind turbines, the turbine, usually used in the horizontal-axial configuration is set on the vertical axis. Wherein classification of wind turbines according to the orientation of their main axes relatively to the horizon is not enough adequate. It is more correct, to divide turbines into orthogonal ones, in which the principal axis is perpendicular airflow and collinear, where the axis is directed along the flow (Fig.2.1).

In connection with the above-mentioned, it appears relevant to make a comparative analysis of energy efficiency of wind turbines in their orthogonal and collinear configurations.

Selecting the configuration (layout) of the wind turbine entails a number of significant constructive and regime features. In the framework of the comparative study of the turbine energy efficiency its blade is considered as a set of segments (local blades) having the same symmetrical sections, each segment is able to change its orientation relatively to direction of absolute airflow, in particular by turning around the own (supporting) axis of the blade.

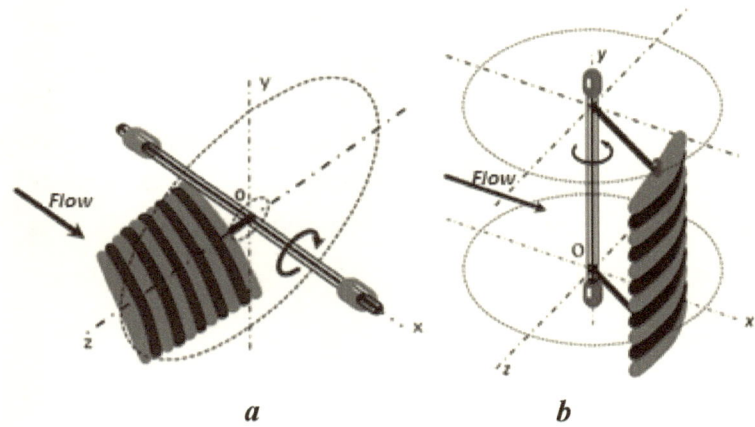

a   b

Fig.2.1. Basic configurations of wind turbines:
a) collinear, OX - the main axis of a turbine, the reference axis of the blade OZ is perpendicular to the main one,
b) orthogonal, OY- the main axis of the turbine, the own (supporting) axis of the blade is parallel to the main one.

If in the orthogonal turbine the orientation of all local blades are generally the same and that allows to form a constructively united blade, then in collinear turbine an adjustable turn of local blades is significantly dependent on their location on the supporting axis. On the other hand effect of airflow on a collinear turbine is in no way connected with turning a blade around the main axis of the turbine while in the orthogonal turbine such impact depends on the position of the axis of the blade and carries an obvious cyclicity.

2.1. The classical theory of an ideal windmill.

According to the theory of N.E.Zhukovsky for the ideal wind turbine [44,45], the limiting index of wind energy using is 0.593. That is the ideal wind turbine with an infinite number of blades can extract 59.3% of the energy of the airflow passing through its transverse section. Actually, in practice, the best maximum value of the index of wind energy using wind plant reaches 0.45, for slow-moving ones it is 0.36-0.38.

In Fig.2.2 an ideal wind turbine is presented as an active air area of flow which has constant cross-section equals to a section of the input channel of the windmill and this channel is orthogonal both its axis and the direction of the flow.

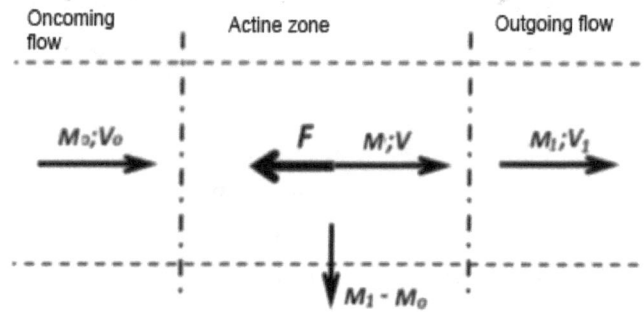

Fig.2.2. Flow diagram in an ideal windmill.

Airflow having at the entrance to the active zone mass flow rate $M_0=(dm/dt)_0$ Kg/sec and the velocity V_0 M/sec, meets counteraction of working bodies of the windmill, presented reduced force F, as a result of which flow dissipation happens. Part of the air leaves the tube of flow and as a result of its mass flow rate and speed respectively is decreased to M_1 and V_1.

In this case, taking into account the incompressibility of air in the flow, Its mass flows and velocities are mutually proportional:

$$M_1/M_0 = V_1/V_0 = d, \tag{2.1}$$

where $d \leq 1$ is the proportionality coefficient determining degree of slowing of the airflow. The quantities M and V are the averaged values of the mass flow rate and airflow rate in the active zone of windmill.

Scattering of airflow overcoming resistance of the working parts when passing through active zone, is connected with the power extracted by the wind wheel from the stream. In accordance with the classical Theorem on kinetic energy, the power produced by airflow is equal to the difference of the kinetic energy in incoming and outgoing flows

$$N = \frac{MV_0^2}{2} - \frac{MV_1^2}{2} \tag{2.2}$$

On the other hand according to the momentum theorem

$$F = \Delta p/\Delta t = MV_0 - MV_1, \tag{2.3}$$

Change in the momentum of the flow is caused by forces acting from the working parts of the windmill. Since the power of the force F, by definition, $N = \Delta A/\Delta t = F\Delta L/\Delta t$, where ΔL is the length of active zone, $\Delta L/\Delta t$ - average flow velocity V in the active zone, the expression for the power becomes

$$N = M(V_0 - V_1)V, \tag{2.4}$$

and the comparison (2.2) and (2.3) gives the value of the mean velocity

$$V = (V_0 + V_1)/2. \tag{2.5}$$

To determine WEI (η), the relation

$$\eta = N / \left(\frac{M_0 V_0^2}{2} \right).$$

The substitution of (2.1), (2.4), and (2.5) into it yields a formula for WEI (Wind Energy Index)

$$\eta = \frac{1}{2}(1 + d)(1 - d^2) \tag{2.6}$$

The maximum WEI is reached at $d = 1/3$ and is

$$\eta_{max} = 16/27 = 0.593.$$

According to the theory of prof. G.Kh.Sabinin, a coefficient of wind energy using in the ideal wind mill is 0.687. The difference is that in determining the axial force of the flow pressure on the wind wheel impulse forces is calculated from the vortex solenoid, in the place where it is accepted the already established cylindrical form, and not in the moment of its formation, as was accepted by the former theories. Since the solenoid in the cylindrical part has the cross-sectional area which is larger than the area swept out with the wind wheel, the axle force and the coefficient of wind energy using according to the theory of G.Kh.Sabinin, are obtained rather large. However, modern international researchs [46] still confirm the correctness of canonical solution of N.E. Zhukovsky.

2.2. The interaction of the flow with the blade wind turbines.

Investigation of the effect of airflow on the symmetric plane-convex blade (Fig.2.3), with an assumption of the superposition of the action of dissimilar applied forces, allows to consider universal problem using as the principle of frontal resistance and Lift Force of the wing [47].

The force of the transverse frontal impact of airflow on the flat blade base [9, 10] is directed along the normal On to the cross-section of the blade (Fig.2.3a) and is

$$F_1 = c_{fn} S_n cos\sigma \frac{\rho V^2}{2}. \qquad (2.7)$$

The corresponding transverse elevating force is absent due to the symmetry of the blade.

With longitudinal interaction (Fig.2.3b), the force of frontal impact is directed along O and is equal to

$$F_2 = c_{f\tau} S_\tau sin\sigma \frac{\rho V^2}{2}, \qquad (2.8)$$

and the lifting force [9, 11] - along On and is equal to

$$F_3 = c_{l\tau} S_\tau sin\sigma \, \rho V^2 / 2. \qquad (2.9)$$

Here $c_{fn}, c_{f\tau}$ - coefficients of drag blades in the transverse (On) and longitudinal ($O\tau$) directions; $c_{l\tau}$ - coefficient of Lift Force; S_n, S_τ – areas of transverse (orthogonal to On) and longitudinal (orthogonal to $O\tau$) section of the blade; σ - is the angle, formed by vector of the relative flow velocity with axis of symmetry On; ρ is the density of air; V -relative air velocity.

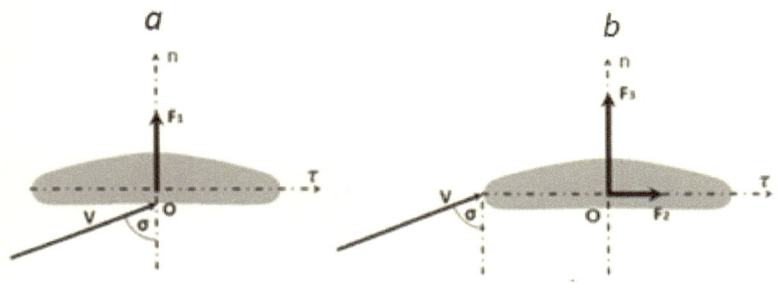

Fig.2.3. Interaction of airflow with a plane-convex blade:
a) transverse; *b*) longitudinal.

The index of the orientation of the segment of the blade of a collinear turbine is the angular coordinate - the angle φ between the direction of the flow Vo and the normal n to the flat base of the local blade (Fig.2.4).

Air velocity relative to the blade is defined as the vector difference of the absolute velocity of flow $\vec{V}o$ and of the portable (circumferential) velocity of the base point of the local blade $\vec{V}e$, in its rotation around the main axis of the turbine:

$$\vec{V} = \vec{V}o - \vec{V}e$$

Taking into account that the vectors of absolute and portable velocities are mutually perpendicular, relative velocity can be calculated from expression

$$V^2 = Vo^2 + Ve^2.$$

If the ratio of absolute and characteristic circumferential velocities is set by the circumferential velocity parameter

$$k = Ve_1/Vo,$$ (2.10)

then the local portable velocity is defined as

$$V_e = V_0 kz,$$ (2.11)

where $z = Z/Z_1$ – specific (relative) coordinate of the blade section. Then the expression for the relative velocity takes the form

$$V^2 = V_0^2(1 + (kz)^2).$$ (2.12)

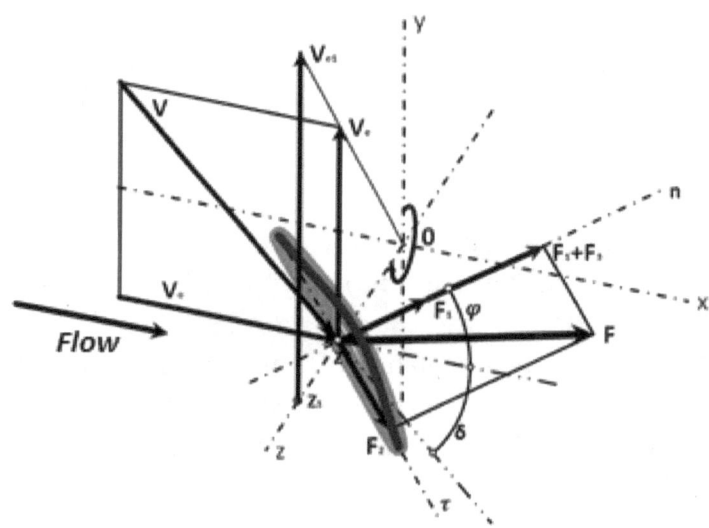

Fig.2.4. Scheme of interaction of airflow with segment (local blade) of collinear wind turbines: Ox - the main axis of the turbine, Oz - the reference axis blades, Z and Z1- base coordinates respectively flowing and final segments of the blade, (OZ$_1$- length of blade), V$_{e1}$ - characteristic circumferenced velocity of blade.

Since the projection of the relative velocity on normal of the blade is equal to the difference of the corresponding projections of absolute and portable velocities

$$V cos(\vec{n}; \vec{V}) = Vo \, cos\varphi - Ve \, sin\varphi,$$

then

$$cos(\vec{n}; \vec{V}) = \frac{cos\varphi - kz \, sin\varphi}{\sqrt{1+(kz)^2}}, \tag{2.13}$$

and applied forces of airflow are converted to the form

$$F_1 = Fo\sqrt{1+(kz)^2}(cos\varphi - kz \, sin\varphi), \tag{2.14}$$

$$F_2 = s_2 Fo\sqrt{1+(kz)^2} \, (sin\varphi + kz \, cos\varphi), \tag{2.15}$$

$$F_3 = s_3 Fo\sqrt{1+(kz)^2} \, (sin\varphi + kz \, cos\varphi), \tag{2.16}$$

Here

$$F_0 = c_{fn} \, S_n \, \rho V_0^2 / 2 \tag{2.17}$$

is the reduced force of the flow, which is the force of the transverse frontal action applied to the fixed plate in the normal direction of flow,

$$s_2 = \frac{c_{f\tau} S_\tau}{c_{fn} S_n} \tag{2.18}$$

is the reduced coefficient of the longitudinal frontal resistance,

$$s_3 = \frac{c_{l\tau} S_\tau}{c_{fn} S_n} \tag{2.19}$$

is the reduced index of Lift force.

Accordingly, the main moment of the forces applied to segment relative to the main axis of the turbine OX

$$M = F_1 Z sin\varphi - F_2 Z cos\varphi + F_3 Z sin\varphi . \tag{2.20}$$

The main moment in the multi-segment blade M_I is determined by summing the moments in the local blades. Using expressions obtained earlier for the local forces F_{1i}, F_{2i} and F_{3i}, it turns out

$$M_I = Z_1 \sum_{i=1}^{I} F_{0i} \, z_i \sqrt{1+(kz_i)^2} \, (((cos\varphi_i - kz_i sin\varphi_i)sin\varphi_i -$$

$$-s_2(sin\varphi_i + kz_icos\varphi_i)cos\varphi_i + s_3(sin\varphi_i + kz_icos\varphi_i)sin\varphi_i),$$

$$(2.21)$$

where it is the current index of the local blade (segment), I - a number of segments in the multi-segment blade, F_{0i} - a reduced force of action on the local blade.

If we assume that the segments are geometrically identical, then the local reduced force is proportional to the length of segment

$$F_{0i} = F_o\Delta z_i,$$

$$(2.22)$$

where

$$F_o = \sum_{i=1}^{I} F_{0i}$$

$$(2.23)$$

is the total reduced force of the flow on the multi-segment lobe, $\Delta z_i = \Delta Z_i/Z_1$ - relative length of the segment. Respectively

$$M_I = F_0Z_1 \sum_{i=1}^{I} \Delta z_i \, z_i\sqrt{1 + (kz_i)^2} \, (((cos\varphi_i - kz_isin\varphi_i)sin\varphi_i -$$
$$-s_2(sin\varphi_i + kz_icos\varphi_i)cos\varphi_i + s_3(sin\varphi_i + kz_icos\varphi_i)sin\varphi_i).$$

$$(2.24)$$

The limiting transition from the final local blade to elementary one, located in the range (z; $z + dz$), allows obtaining a universal integral relation

$$M_z = FoZ_1 \int_0^z z\sqrt{1 + (kz)^2} \, ((cos\varphi - kz \, sin\varphi)sin\varphi -$$
$$-s_2(sin\varphi + kz \, cos\varphi)cos\varphi + s_3(sin\varphi + kz \, cos\varphi)sin\varphi)dz,$$

$$(2.25)$$

suitable for describing both a multi-segment blade and blade with a base in the form of a continuous line surface.

The amount of energy extracted from interacting airflow by blade is equal to work of forces applied to the blade on the displacement of this blade or the work of the moment of these forces on the corresponding angular displacement. By the full turn of the turbine, work is $A_{2\pi} = 2\pi M_z$.

When calculating energy, as a scale it is advisable to consider effective cyclic air energy $A_\vartheta = 2\pi FoZ_I$, which is equal to work of the reduced force of action of the flow on a displacement of the end of the blade for the full revolution (cycle) of the turbine. Relevant coefficient of energy intake ($a_a = A_a/A_9$)

$$a_z = \int_0^z z\sqrt{1 + (kz)^2} \, ((\cos\varphi - kz\sin\varphi)\sin\varphi -$$

$$-s_2(\sin\varphi + kz\cos\varphi)\cos\varphi + s_3(\sin\varphi + kz\cos\varphi)\sin\varphi)dz. \quad (2.26)$$

If for the scale of the angular momentum M_z to take a moment of reduced force Fo on the arm Z_1, then the expression (2.25) also reduces to (2.26), that is, the quantity a_z acquires meaning, simultaneously, as a cyclic coefficient of extraction of energy from airflow and as the principal relative moment of forces of airflow applied to the blade.

Similarly, the calculation of extraction of a power of airflow is carried out. The power extraction N_z is equal to the product of principal moment and the angular velocity V_{e1}/Z_1

$$N_z = M_z V_{e1}/Z_1. \quad (2.27)$$

The choice $N_9 = FoVo$ as the scale of the reduced (effective) power of airflow gives simple relation connecting cyclic values indexes of extraction of energy and power of airflow

$$n_z = ka_z. \quad (2.28)$$

Accordingly, the absolute value of the power extraction of airflow

$$N_z = n_z FoVo. \quad (2.29)$$

The orthogonal turbine rotates around the main shaft (axis) Oy and is equipped with a blade mounted on the own shaft (axis) parallel to the main shaft and located on the axis of symmetry of the blade (Fig.2.5).

Own blade shaft has a drive, turning the blade to regulate it interaction with the flow. Taking into account that vectors of absolute and portable velocities form an angle ($\pi/2 + \alpha$), the relative velocity can be calculated by the cosine theorem

$$V^2 = Vo^2 + Ve^2 - 2VoVe\cos(\pi/2 + \alpha). \quad (2.30)$$

If the ratio of absolute and portable flow rates is set by the circumferential velocity parameter

$$k = Ve/Vo, \tag{2.31}$$

then

$$V^2 = V_0^2(1 + k^2 + 2k \sin\alpha). \tag{2.32}$$

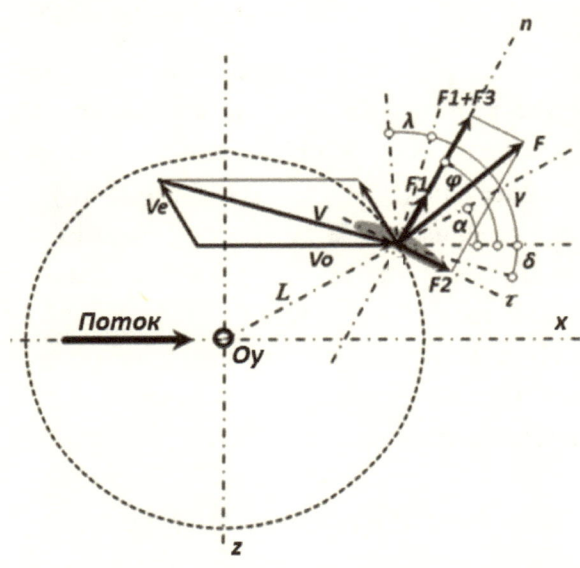

Fig.2.5. Scheme of an interaction of airflow with a blade
of an orthogonal wind turbine: Oy is the main axis
of a turbine, L - rotor radius. Angles of orientation:
turbines- α, blades - φ.

Since the projection of relative velocity on the normal of the blade is equal to a difference of the corresponding projections of absolute and portable velocities

$$V\cos(\vec{n}; \vec{V}) = Vo\,\cos(\varphi) - Ve\,\sin(\varphi - \alpha),$$

then angle $\sigma = (n; V)$ is determined from relation

$$\cos(\vec{n}; \vec{V}) = \frac{\cos\varphi - k\sin(\varphi - \alpha)}{\sqrt{1 + k^2 + 2k\sin\alpha}}. \tag{2.33}$$

The applied forces of airflow are converted to the form

$$F_1 = Fo\sqrt{1 + k^2 + 2k\sin\alpha}\ [\cos\varphi - k\sin(\varphi - \alpha)], \tag{2.34}$$

$$F_2 = s_2 Fo\sqrt{1 + k^2 + 2k\sin\alpha}\ [\sin\varphi + k\cos(\varphi - \alpha)], \tag{2.35}$$

$$F_3 = s_3 Fo\sqrt{1 + k^2 + 2k\sin\alpha}\ [\sin\varphi + k\cos(\varphi - \alpha)], \quad (2.36)$$

Accordingly, the main moment of the applied forces

$$M_\alpha = F_1 L\sin(\varphi - \alpha) - F_2 L\cos(\varphi - \alpha) + F_3 L\sin(\varphi - \alpha). \quad (2.37)$$

The amount of energy extracted from the airflow interacting with the blade is equal to work of the forces applied to the blade on the displacement of this blade, or the work of the moment of these forces on the corresponding angular displacement

$$A_\alpha = \int_0^\alpha M_\alpha d\alpha,$$

Taking into account earlier expressions for current values of the forces F_1, F_2 and F_3,

$$A_\alpha = FoL \int_0^\alpha \sqrt{1 + k^2 + 2k\sin\alpha}\ (((\cos\varphi - k\sin(\varphi - \alpha))\sin(\varphi - \alpha) -$$
$$-s_2(\sin\varphi + k\cos(\varphi - \alpha)]\cos(\varphi - \alpha) + s_3(\sin\varphi + k\cos(\varphi - \alpha))\sin(\varphi - \alpha))d\alpha.$$
$$(2.38)$$

When calculating energy as a scale it is advisable to consider *effective cyclic air energy* $A_э = 2\pi FoL$, equal to the work of reduced force of the flow on moving the axis blade for the full revolution (cycle) of the turbine. Relevant coefficient of energy extraction ($a_\alpha = A_\alpha / A_э$)

$$a_\alpha = \frac{1}{2\pi} \int_0^\alpha \sqrt{1 + k^2 + 2k\sin\alpha}\ (((\cos\varphi - k\sin(\varphi - \alpha))\sin(\varphi - \alpha) -$$
$$-s_2(\sin\varphi + k\cos(\varphi - \alpha))\cos(\varphi - \alpha) + s_3(\sin\varphi + k\cos(\varphi - \alpha))\sin(\varphi - \alpha))d\alpha.$$
$$(2.39)$$

Similarly, the calculation of cyclic power extraction of the airflow. Average selection of capacity per revolution of the turbine N_0 is equal to the ratio of corresponding energy A_0 to the turbine revolution time $2\pi L/kVo$

$$N_0 = \frac{FoVo}{2\pi} k \int_0^\alpha \sqrt{1 + k^2 + 2k\sin\alpha}\ (((\cos\varphi - k\sin(\varphi - \alpha))\sin(\varphi - \alpha) -$$
$$-s_2(\sin\varphi + k\cos(\varphi - \alpha))\cos(\varphi - \alpha) + s_3(\sin\varphi + k\cos(\varphi - \alpha))\sin(\varphi - \alpha))d\alpha.$$
$$(2.40)$$

The choice of the scale of the reduced (effective) power of airflow $Ne = FoVo$ gives simple relation connecting cyclic values of energy intake indexes and power of airflow

$$n_0 = ka_0. \quad (2.41)$$

Accordingly, the absolute value of the power extraction of the airflow averaged over one revolution

$$N_0 = n_0 FoVo. \qquad (2.42)$$

2.3. Optimization model of wind turbine.

Energy-efficient functioning the *collinear* wind turbine assumes a certain shape of the blade based on the optimal location of its segments (elements) in relation to the airflow.

Study of the dependence of $a(z)$ about the extremum gives an opportunity to identify the optimal orientation of elements blades of a collinear wind turbine, depending on their position on the reference axis, that is, the optimal dependence $\varphi=\varphi(z)$. The extremum condition $da_z/dz = 0$ gives the initial relation

$$\frac{d}{dz}((cos\varphi - kz\,sin\varphi)sin\varphi - s_2(sin\varphi + kz\,cos\varphi)cos\varphi +$$

$$+s_3(sin\varphi + kz\,cos\varphi)sin\varphi) = 0, \qquad (2.43)$$

which after differentiation and subsequent transformations becomes

$$tg(2\varphi) = \frac{1+kz\frac{s_3}{1-s_2}}{kz-\frac{s_3}{1-s_2}} \qquad (2.44)$$

In order to represent the relation $\varphi=\varphi(z)$ in an explicit form, the triangle formed by vectors of absolute, relative and transport velocities is considered (Fig.2.4), from which follows

$$kz = Ve/Vo = tg\delta. \qquad (2.45)$$

If, in addition, introduce the angle of "the additional optimal rotation" λ, characterizing the longitudinal effect of airflow, which is calculated from the dependence

$$tg\lambda = \frac{s_3}{1-s_2}, \qquad (2.46)$$

then the comparison of the expressions for tg(2β), tgδ and tgλ gives a formula

$$tg(2\varphi) = 1/tg(\delta - \lambda). \qquad (2.47)$$

Accordingly in explicit form (2.44) becomes

$$\varphi = (\pi/2 - \delta + \lambda)/2. \qquad (2.48)$$

Optimal blades configuration of collinear turbine (Fig.2.6) depends on its speed regime.

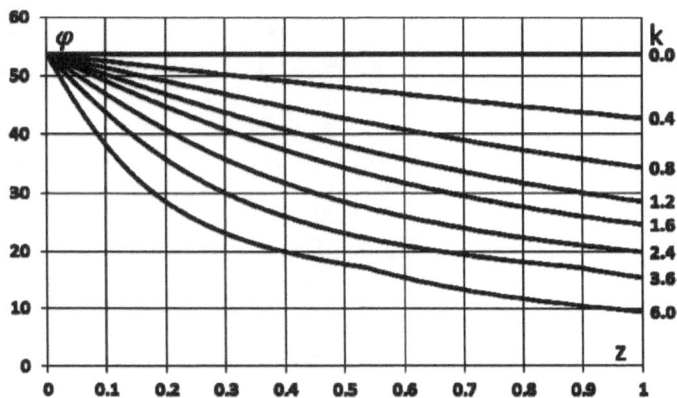

Fig.2.6. Indicators of the optimal orientation
of Collinear Wind Turbine blades: $s_2 = 0.04$; $s_3 = 0.35$;
$k = 0.0 \div 6.0$. z - is the specific length of the blade.

If at essentially low peripheral velocities ($k \rightarrow 0$), the optimal blade is flat or plano-convex (if there is Lift Force), then the growth of k leads to increasing of mutual angular displacement of the blade elements. In this case, the flat base of the blade turns into a linear surface. Behavior of functions (k) with a significant increase in circumferential velocity

$$\lim_{k \to \infty} \varphi = \pi/2,$$

reflects the fact of asymptotic approximation of the optimal blade to the position of the plane which is perpendicular to the main axis of the turbine.

Operation of the *orthogonal* wind turbine is possible only with a certain arrangement of blades relative to the airflow. Optimization model analysis of blade interaction with airflow allows you to determine the blade kinematics, ensuring maximum energy release of flow.

The study of dependence $a_0(\varphi)$ by extremum gives a possibility to identify the optimal orientation of blades of the orthogonal wind turbine, depending on their axes, that is, determine the optimal relation $\varphi=\varphi(\alpha)$. The extremum condition $da_\alpha/d\varphi = 0$, gives the initial relation

$$\frac{d}{d\varphi}(((cos\varphi - ksin(\varphi - \alpha))sin(\varphi - \alpha) - s_2(sin\varphi + kcos(\varphi - \alpha))cos(\varphi - \alpha) +$$

$$+s_3(sin\varphi + kcos(\varphi - \alpha))sin(\varphi - \alpha)) = 0, \tag{2.49}$$

which after differentiation and subsequent transformations becomes

$$tg(2\varphi - \alpha) = \frac{\frac{1+ksin\alpha}{kcos\alpha} \pm \frac{s_3}{1-s_2}}{1 \mp \frac{1+ksin\alpha}{kcos\alpha}\frac{s_3}{1-s_2}} \tag{2.50}$$

To solve this trigonometric equation, we can represent the relation $\varphi=\varphi(\alpha)$ in an explicit form, for what we consider the triangle formed by the vectors of absolute, relative and transport velocities (Fig.2.5), applying to it the sine theorem

$$\frac{Ve}{sin(\vec{V};\vec{V}o)} = \frac{Vo}{sin(\vec{V};\vec{V}e)}.$$

Converting this expression to the form

$$k = Ve/Vo = sin\delta/sin(\pi/2-\alpha-\delta),$$

it is possible to determine the angle δ characterizing the angular deviation of the relative airflow

$$tg\delta = kcos\alpha/(1+k\ sin\alpha). \tag{2.51}$$

The reverse value of $(1+k\ sin\alpha)/kcos\alpha$ in the expression for $tg(2\varphi - \alpha)$ characterizes the tangent of the angle which complements the angle δ to the right angle ($\gamma = \pi/2 - \delta$ in Fig.2.5)

$$tg\gamma=(1+k\ sin\alpha)/kcos\alpha. \tag{2.52}$$

If you consider the angle of "additional optimal rotation" λ, characterizing the effect of Longitudinal action of the airflow, which Is calculated from the dependence

$$tg\lambda = \frac{s_3}{1-s_2}, \tag{2.53}$$

then the comparison of the expressions for tg(2β-α), tgγ and tgλ gives a formula

$$tg(2\varphi - \alpha) = tg(\gamma \pm \lambda).$$

Accordingly, in an explicit form, the solution α (Fig.2.7) takes the form

$$\varphi = (\alpha + \gamma \pm \lambda)/2. \tag{2.54}$$

Here the angles α, γ, and λ are presented their main values in the range from -π / 2 to π / 2, in particular:

$$\gamma = arctg((1+ksin\alpha)/kcos\alpha), \tag{2.55}$$

$$\lambda=arctg(s_3/(1-s_2)), \tag{2.56}$$

at that, the value of the angle λ is added (+) when a blade is found in the first or fourth quadrants (α from -π/2 to π/2) and subtracted (-) in the second and third quadrants.

Comparison of optimal blade configurations collinear wind turbines with orthogonal ones (Fig.2.6, 2.7), shows a significant difference in the approaches to their optimization.

In a collinear turbine, the optimal orientation of blade does not depend on the rotation of the turbine, there is no effect of inversion, but the local orientation of the blade is associated with the distance Z of the blade element (section) from the main axis of the turbine. Proceeding from this the optimum blade is formed not on a flat, but on linear base, the shape of which, in addition, depends on the peripheral velocity of the turbine (parameter k).

In the orthogonal turbine, on the contrary, the optimal orientation of the blade does not change along its axis, that is own configuration of the blade is unchanged - flat or plane-convex. But the optimal orientation of the blade significantly varies depending on the rotation speed of the turbine around the main axis (values of angle α).

It is necessary to pay special attention to the effect of blade inversion with its optimal orientation - the discontinuous changes in the orientation angle φ at the values of α which are equal 90^0 and 270^0 (-90^0). This effect is typical for un-symmetry "one-sided" blades and requires special technical means for the implementation of such special modes of motion.

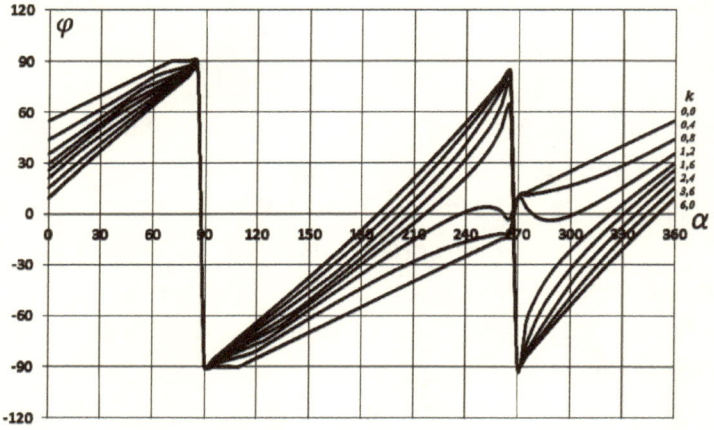

Fig.2.7. Indicators of the optimal orientation
of the orthogonal wind turbine blade: $s_2 = 0.04$;
$s_3 = 0.35$; $k = 0.0 \div 6.0$; α - angle of rotation
of the turbine

Thus, for optimization of orthogonal turbine special permanent actuator is required for blade orientation during the cycle (revolution) around the main axis of the turbine.

2.4. Optimizing of the wind turbine ideal blade.

The ideal blade is understood as a blade in which there are no longitudinal effects of interaction with airflow - as frontal resistance and Lift Force: $s_2 = 0$; $s_3 = 0$. Such object is sufficiently simple to apply methods of exact "non-numerical" analysis, allows you to explore a number of effects that can be subsequently transferred to more complex "non-ideal" objects.

The ideal blade of a *collinear* wind turbine is formed by linear surfaces having straight sections in planes perpendicular to the base axis.

Due to the lack of Lift Force effect the angle of additional rotation of the blade $\lambda=0$, and the optimal orientation of the blade is determined by the relation

$$\varphi = (\pi/2 - \delta)/2. \tag{2.57}$$

Optimal arrangement of the blade in the longitudinal direction τ is determined by the bisector of the angle formed by relative and transport velocity vectors of airflow in each section of the blade (Fig.2.8).

With the optimal orientation of the ideal blade the corresponding maximum index of extract of flow energy

$$a_z = \frac{1}{2}\int_0^z z\sqrt{1 + (kz)^2}\,(\cos\delta - kz(1 - \sin\delta))dz. \tag{2.58}$$

Taking into account that, according to (2.45),

$$\cos\delta = \frac{1}{\sqrt{1 + (kz)^2}}, \quad \sin\delta = \frac{kz}{\sqrt{1 + (kz)^2}},$$

the solution of the integral (2.58) takes the form

$$a_z = \frac{1}{16k^2}\left(kz\left(2kz(2 + (kz)^2) - (1 + 2(kz)^2)\sqrt{1 + (kz)^2}\right) + \ln\left(kz + \sqrt{1 + (kz)^2}\right)\right) \tag{2.59}$$

The corresponding calculation results are presented in Fig.2.9. The total accumulated value of extract coefficient is determined for a collinear turbine by the substitution $z = 1$

$$a_1 = a_{z=1}. \tag{2.60}$$

Limit analysis of the solution (2.58) shows that during rotation of a collinear wind turbine with significantly small circumferential velocities ($k \to 0$), when the relative airflow almost coincides with the absolute ($\delta \to 0$), a ratio of the integral work to the cyclic effective energy takes on value

$$\lim_{k \to 0} a_1 = 1/4. \tag{2.61}$$

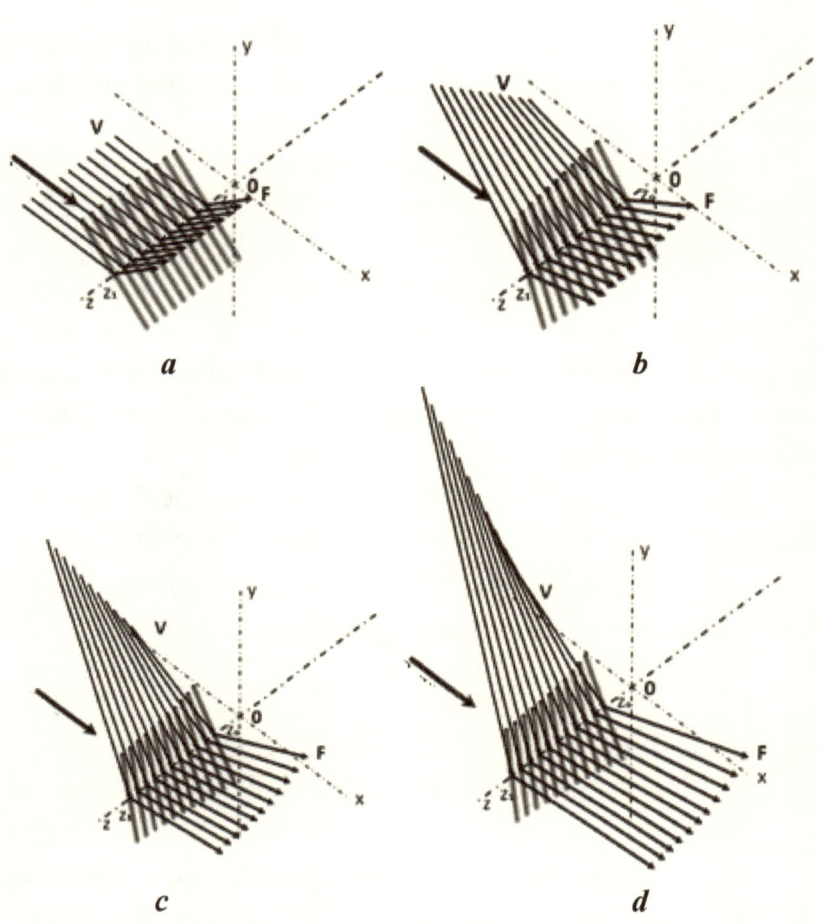

Fig.2.8. Distributed scheme of optimal configuration
of collinear wind turbine blades. Ideal blade.
The direction of the airflow is along the axis OX.
The results of the calculation: *a*) k=0; b) k=1; c) k=2; d) k=3.

Another limiting value of integral work (energy extraction) is related to the behavior of the *a(k)* dependence at significant growth of the circumferential velocity ($k \rightarrow \infty$)

$$\lim_{k \to \infty} a_1 = 1/8. \tag{2.62}$$

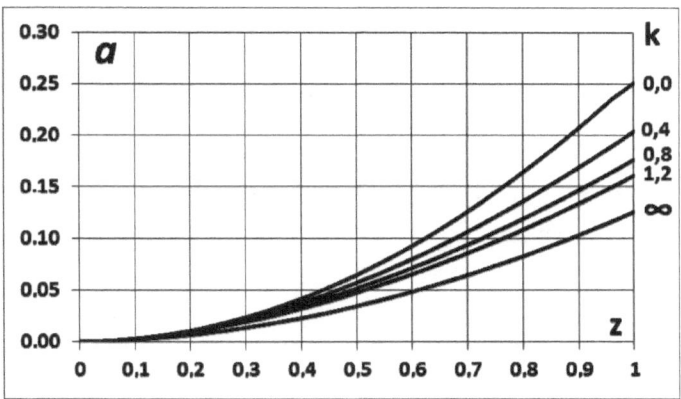

Fig.2.9. Current accumulated values of the extract energy
coefficient of airflow in a collinear wind turbine.
Ideal blade, z is a specific length of blades. Values
of the circumferential velocity parameter $k = 0 \div \infty$.

In *orthogonal* turbine in the absence of Lift Force effect,
angle of additional rotation of the blade $\lambda=0$ and the optimal
orientation of the blade is determined by relation

$$\varphi = (\gamma+\alpha)/2. \qquad (2.63)$$

Optimal arrangement of the blade in the longitudinal
direction Oτ is also determined by the bisector of the angle,
formed by relative and portable velocity vectors of airflow. It is
characteristic that the rotational moment of the force F which is
orthogonal to plane blade base at any rotor position does not take
negative values.

The picture of the optimal orientation of the blades is
significantly depended from speed mode of turbine (Fig.2.10).

If at low circumferential velocities ($k<1$), the mode "wind
vane", when the blade is installed in a parallel position to airflow,
occurs only at $\alpha=90^0$, then for $k>1$ this mode extends to $\alpha=270^0$
(-90^0).

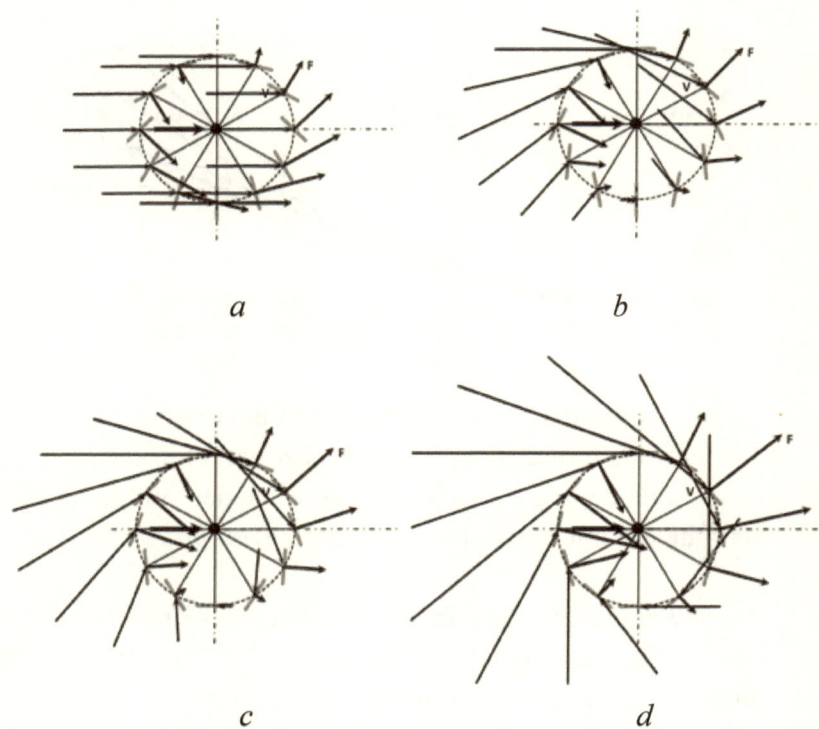

a   *b*

c   *d*

Fig..2.10. Distributed scheme of optimal blades
orientation of orthogonal turbine. The ideal blade:
$s_2 = 0$; $s_3 = 0$; *a*) $k = 0.0$; *b*)$k = 0.8$; *c*) $k = 1.2$; *d*) $k = 2.0$

Furthermore, the behavior of the function $\varphi(\alpha)$ with a significant growth of the circumferential velocity - $\lim_{k\to\infty} \varphi = \alpha$, reflects the fact of the asymptotic approaching the blade to the tangential position in any point of the turbine.

Taking into account the optimal orientation of the ideal blade, the corresponding maximum extract factor of flow energy

$$a_\alpha = \frac{1}{2\pi} \int_0^\alpha \sqrt{1 + k^2 + 2k \sin\alpha} \left(cos\frac{\gamma+\alpha}{2} - ksin\frac{\gamma-\alpha}{2}\right) sin\frac{\gamma-\alpha}{2} d\alpha. \quad (2.64)$$

When the blade rotates around the axis of the wind turbine with essentially small circumferential velocities ($k\to0$), when the relative airflow practically coincides with absolute one ($\delta\rightarrow0$), the integral of the work related to cyclic effective energy, becomes

$$a_{\alpha 0} = \frac{1}{2\pi}\int_0^\alpha (1 - \sin\alpha)d\alpha, \tag{2.65}$$

and the corresponding decision

$$a_{\alpha 0} = \frac{\alpha+\cos\alpha-1}{4\pi}. \tag{2.66}$$

Another limiting configuration of the integral of work (energy extraction) is related to the behavior of the $a(k)$ dependence at significant growth of the circumferential velocity ($k\rightarrow\infty$)

$$a_{\alpha\infty} = \lim_{k\rightarrow\infty} a_{\alpha 0} = \frac{1}{2\pi}\int_0^\alpha \frac{\cos^2\alpha}{4} d\alpha. \tag{2.67}$$

The corresponding solution has the form

$$a_{\alpha\infty} = \frac{\alpha+\frac{\sin 2\alpha}{2}}{16\pi}. \tag{2.68}$$

A family of characteristics that determine current values of the energy extraction of the airflow per revolution of a wind turbine (Fig.2.11), is located between represented above by the limiting configurations.

Dependence of the total accumulated value of extraction coefficient a_∞ on the circumferential velocity parameter k both in the collinear and in the orthogonal turbine has an expressed asymptotic character (Fig.2.12). In particular in a collinear turbine the initial value of energy extraction coefficient $a_0 = 0.25$ (see 2.61), and asymptotic - $a_\infty= 0.125$ (see 2.62).

In orthogonal turbine the value of this characteristic us determined by the substitution $\alpha=2\pi$. An initial value of power extraction coefficient for a full revolution of an orthogonal turbine as $k\rightarrow0$ is $a_0 = 0.5$. At the limit, as $k\rightarrow\infty$, the extraction of the cyclic energy asymptotically tends to the universal value $a_\infty= 0.125$.

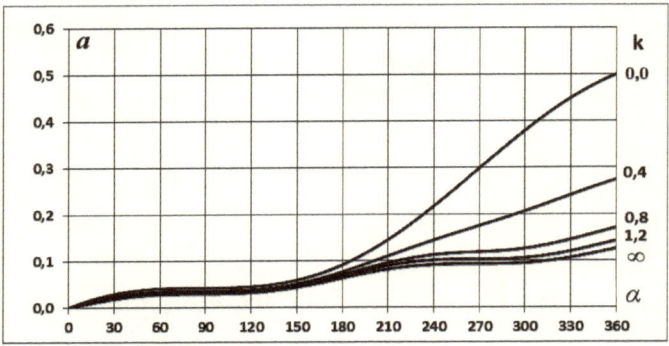

Fig.2.11. Current accumulated values of the extraction energy coefficient of the airflow per revolution of the orthogonal wind turbines. Ideal blade. The values of the circumferential velocity parameter $k=0\div\infty$.

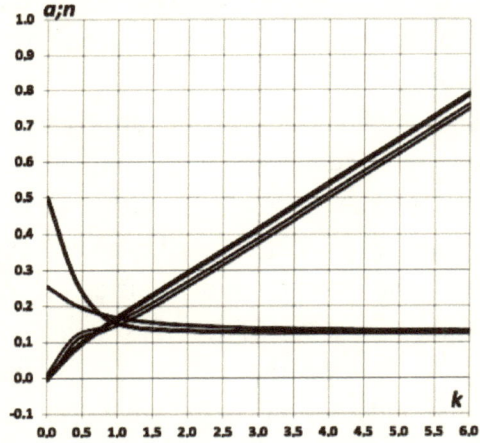

Fig.2.12. Influence of the peripheral velocity of the turbine on the force moment - the extraction of energy per revolution a (thin lines), and extraction of power n (thick lines) of the airflow by the perfect blade. The collinear turbine is marked by ordinary lines, the orthogonal - by double lines.

Since the average torque of airflow per revolution of the turbine is equal to the ratio of the work of these forces A_0 to an angle of rotation 2π and the magnitude of the force moment is FoL value, the energy extraction coefficient of airflow a simultaneously has the meaning of specific (dimensionless) of the torque averaged over the revolution of force moment during influence the flow on the blade of the wind turbine.

Comparison of the limiting values a for collinear and orthogonal turbines shows that if at small circumferential velocities force moment in the collinear turbine turns out to be twice less than in orthogonal one, the limiting values of the angular momentum with increasing the circumferential velocity parameter is stabilized at universal level of extraction of 12.5% of the energy of airflow, regardless of the type of turbine.

Since the power take-off is proportional to the selection energy $(n=k\alpha)$, simultaneously with the stabilization of the moment linear increase in the extraction of cyclic power is observed (Fig.2.12), that is, with increasing peripheral velocity of wind turbine there is a significant increase in its energy efficiency.

2.5. Optimization of non-ideal blade of wind turbines.

The non-ideal flat blade is characterized by the presence of longitudinal frontal resistance of the flow $(s_3 = 0)$. In the absence of Lifting Force $(s_3 = 0)$. Additional rotation of λ is excluded on optimizing the position of the blade, so the optimal orientation of such blade is not different from ideal one.

When calculating the energy characteristics non-ideal blades, in order to reduce labor intensity of analysis, it is advisable to use approximate numerical methods instead of exact analytical methods. Thus, the integral of energy extracted of airflow can be calculated by steps with a high degree of accuracy on a base of the following calculate dependencies.

For collinear turbines

$$a_z = \frac{1}{I} \sum_{i=1}^{z_i I+1} z_i \sqrt{1 + (kz_i)^2} \, (((cos\varphi_i - kz_i sin\varphi_i)sin\varphi_i -$$
$$-s_2(sin\varphi_i + kz_i cos\varphi_i)cos\varphi_i + s_3(sin\varphi_i + kz_i cos\varphi_i)sin\varphi_i),$$

(2.69)

where I is the number of calculated steps. Calculation of power take-off coefficient is realized as $n_z = ka_z$.

For orthogonal turbines

$$a_\alpha = \frac{1}{I} \sum_{i=1}^{\alpha I/2\pi} \sqrt{1 + k^2 + 2k\,sin\,\alpha_i} \, \{[(cos\varphi_i - ksin(\varphi_i - \alpha_i)]sin(\varphi_i - \alpha_i) -$$
$$-s_2[sin\varphi_i + kcos(\varphi_i - \alpha_i)]cos(\varphi_i - \alpha_i) + s_3[sin\varphi_i + kcos(\varphi_i - \alpha_i)]sin(\varphi_i - \alpha_i)\},$$

(2.70)

where I is the number of calculated steps. Calculation of the power take-off coefficient is realized as $n_\alpha = ka_\alpha$.

The presence of resistance always leads to displacement of the resultant force $\vec{F} = \vec{F_1} + \vec{F_2}$ at a direction which is opposite to the rotation of the turbine. In such case, the value of torque in the turbine is reduced. In orthogonal turbine, in addition, sectors with negative values of the rotational moment appear (Fig.2.13), the boundaries of which are expanded with the growth of resistance parameter s_2.

Developed on the basis of the optimization model calculation methods allow significantly to reduce labor intensity study wind turbines, through the implementation of comprehensive and large-scale numerical experiments.

The results of such studies are summarized in tables, nomograms, quasi-empirical dependencies and are used in studying, calculating and designing of wind power plants.

For example, when evaluating the effect of a longitudinal frontal resistance and Lift Force on the characteristics energy and power extracted from the airflow is noted significant differences for collinear and orthogonal turbines.

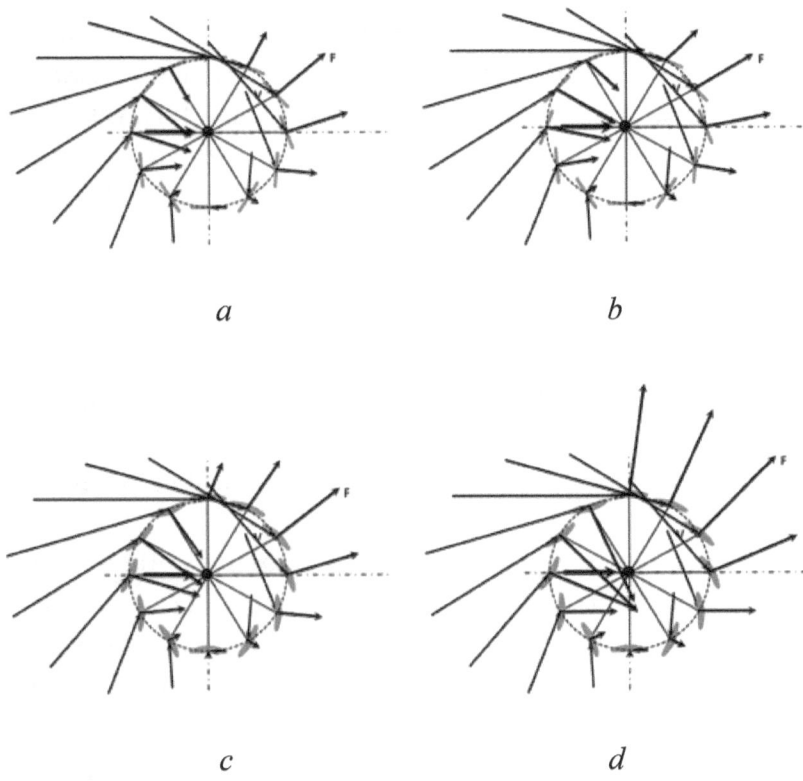

a   *b*

c   *d*

Fig.2.13. Distributed scheme of optimal blades
orientation of the orthogonal turbine: $k = 1.2$.
Non- ideal flat blade: $s_3=0.0$; *a)* $s_2=0.01$; *b)* $s_2=0.05$.
Plane-convex blade: $s_2=0.04$; *c)* $s_3=0.1$; *d)* $s_3=0.3$.

The application of numerical calculation techniques allows identifying the significant effect of the longitudinal frontal resistance to the characteristics of energy (power) extraction from the airflow. In turbines with non-ideal flat (ruled) blades power extraction characteristics become pronounced maximum (Fig.2.14a,b), and in orthogonal turbines (double lines in Fig.2.14), even with a substantially small longitudinal resistance by two orders of magnitude smaller than transverse one (s_2=0.01} the cyclic fraction of the energy extraction from flow does not usually reach 20%. In this way, the energy efficiency of a real (imperfect) flat blade in an orthogonal turbine is substantially limited.

Power extraction characteristics of collinear turbines also have a pronounced maximum, however, collinear turbines (solid lines in Fig.2.14) as for energy efficiency are much higher than orthogonal ones, especially at high-speed modes, where orthogonal turbines are sometimes generally unable to extract the energy of airflow.

Partial compensation of such negative effect can be achieved by using Lift Force effect due to the use of a wing-like flat-convex (linear-convex) blade (Fig.2.14c,d). Noticeable growth of torque is achieved due to, first, additional turn of the blade, while optimizing its position in the direction of rotation of the turbine by angle $\gamma/2$, and, second, the superposition of the force of the transverse frontal effects F_1 with lifting force F_3.

Even with a significant longitudinal resistance of flow (s_2=0.04), the blade-wing allows using orthogonal turbine on efficient modes with extraction more than half the power of the air flow power (Fig.2.14d), however, at the same time, it is saved the extreme nature of relation "power-peripheral speed", limiting the possibility of working on a substantially high revving speeds.

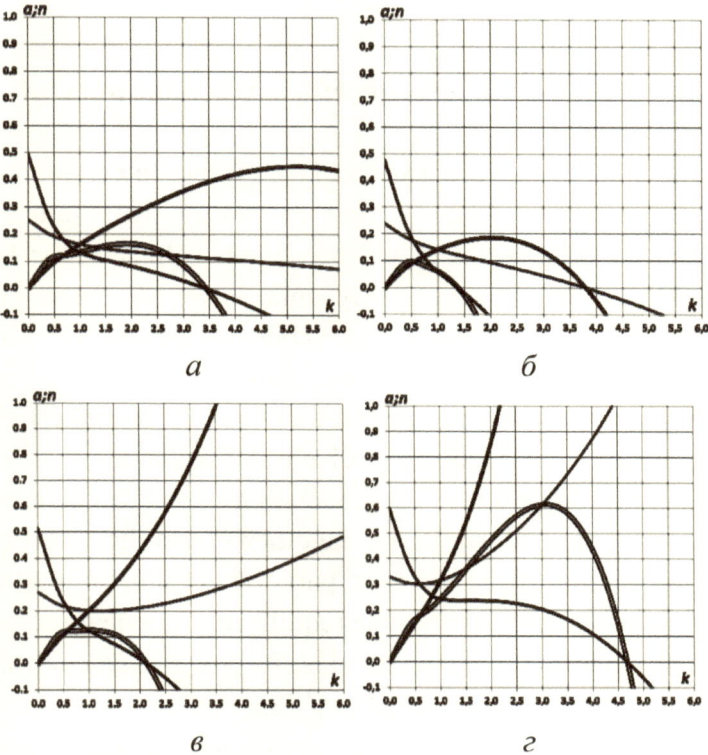

Fig.2.14. Influence of the peripheral speed of the turbine on the force moment (energy extraction) *a* (fine lines) and power extraction *n* (thick lines) of the airflow. Non-ideal blade. Collinear turbine - conventional lines, orthogonal - double lines. Linear blade, $s_3=0$; *a*) $s_2=0.01$; *б*) $s_2=0.05$. Linear-convex blade, $s_2=0.04$; *в*) $s_3=0.1$; *г*) $s_3=0.3$

This effect is removed when used on collinear turbines with vane linear-convex blades providing continuous monotonic growth of the specific power at any values of the rotation speeds of turbines.

2.6. Equilibrium energy efficiency indicators of wind engines.

The theory of an ideal wind plant assumes a certain ratio of axial and rotational forces acting on blades. Only under this condition, maximum power will be obtained. In fact, the blades are working by their own aerodynamic laws. Mismatch energy requirements of wind plants and aerodynamic power of the blades reduce the index of wind energy used η (WEI).

In evaluation and comparative analysis of energy efficiency of wind plants, it is necessary to have the technique that allows for mutual reconciliation of parameters: on the one hand, from the energy model ideal windmill, on the other - from the optimization model interaction of the working parts of the windmill with the airflow, and obtaining compatible (equilibrium) indicators.

In the theory of ideal wind plant, its power is calculated as

$$N_\eta = \eta M_0 V_0^2 / 2, \qquad (2.71)$$

Mass expenditure of air through the active zone of wind plant

$$M_0 = \rho S_0 V_0 , \qquad (2.72)$$

where S_0 – the area of an orthogonal section of the active zone, calculated as the cross-sectional area of the wind wheel (Wind turbine), which is perpendicular to the airflow. When this expression is substituted, taking into account (2.6), the corresponding calculate dependence is obtained

$$N_\eta = (1 + d)(1 - d^2) S_0 \rho V_0^3 / 4. \qquad (2.73)$$

On the other hand, according to (2.29) and (2.42), the power extracted by K blades of the wind wheel (rotor) from the stream of air

$$N_n = nKFoVo. \qquad (2.74)$$

Given that the value of the reduced force acting on a single blade in accordance with (2.17),

$$F_0 = c_{fn} S_n \rho V^2 / 2,$$

where $V = V_0(1+d)/2$ is the average flow velocity in an active zone according to (2.1) and (2.5), the total power acting on the blade of force is

$$N_n = n c_{fn} K S_n \rho V_0^3 (1 + d)^3 / 16. \qquad (2.76)$$

Comparison (equating) the power values N_n and N_η gives the equilibrium value of the flux scattering coefficient

$$d^* = (4S_0 - nc_{fn}KS_n)/(4S_0 + nc_{fn}KS_n) , \qquad (2.76)$$

and the subsequent substitution in (2.6) - a reduced (equilibrium) value of WEI (η_t).

In addition, taking as a basis the expression (2.3) for the reduced force of the wind plant pressure on the airflow (resistance to flow) F, and substituting in it the mean mass flow rate value

$$M = (M_0 + M_1)/2 = M_0(1+d)/2, \qquad (2.77)$$

Taking into account that the loss of pressure of the flow is determined by the ratio of the resistance force F to the area of section of the flow S_0, it is possible to obtain the calculated dependence for pressure loss

$$f = (1 - d^2)\rho V^2 / 2. \qquad (2.78)$$

Comparison of the results of calculation of collinear and orthogonal wind plants (Tables 1, 2 and 3), with identical dimensions of wind turbines and identical speed parameters, reveals the following circumstance.

Although in the considered ranges of values of geometric parameters and mode parameters the collinear rotor is more effective from the point of view of energy extraction from a unit area of the flow section, the absolute power of wind turbines is practically the same because of the geometric features of their design.

Table 2. An approximate calculation of the collinear wind power plant.

Parameter	Description	Formula	Value
The density of air, Kg / m	ρ	Relative constant	1,20
Coefficient of frontal resistance	cfn	Relative constant	1,33
The efficiency of the generator, 0 … 1	η_g	Relative constant	0,870
Efficiency of electronic inverter 0 … 1	η_i	Relative constant	0,850
Limited index of wind energy using of wind plant	ηmax	Constant	0,593
Coefficient of circumferential speed	k	Specified value	2,00
Wind speed, M / s	Vo	Specification	11
External diameter Rotor, m	D	Specification	3,00
Rotor Height (Blades), m	H	Specification	-
The width of the blade	B	Specification	1,50
Number of blades.	K	Specification	3
Area of blade	Sn	$Sn=DB/2$	2,25
Streamlined area of wind plant, m^2	So	$So=\pi D^2/4$	7,07
Mass flow Air through Streamlined Area, kg / sec	Mo	$Mo=\rho SoVo$	93
Power of Airflow W	E	$E=MoVo^2/2$	5642
Power takeoff coefficient of flow	n	Calculation model	0,60
Equilibrium coefficient of flow deceleration	d*	$d*=(4So-nc_{fn}KSn)/(4So+nc_{fn}KSn)$	0,680
Index of air energy using of wind plant	η_t	$\eta_t=(1+d*)(1-d*^2)/2$	0,452
Efficiency of wind plant	η	$\eta=\eta g\eta i\eta t$	0,334
Power of plant, kW	N	$N=E\eta/1000$	1,88
Specific power of wind plant, kW / m^2	n_s	$n_s=N/So$	0,27
Loss of flow pressure in wind engine, Pa	f	$f=(1-d*^2)\rho Vo^2/2$	39,05

Table 3. An approximate calculation of the orthogonal wind power plant.

Parameter	Description	Formula	Value
Density of air, kg / m³	ρ	Relative constant	1,20
Coefficient of frontal resistance	c_{fn}	Relative constant	1,33
Efficiency of generator, 0 ... 1	η_g	Relative constant	0,870
Efficiency of electronic inverter 0 ... 1	η_i	Relative constant	0,850
Limit of Index of air energy using of wind plant	ηmax	Relative constant	0,593
Coefficient of circumferential Speed	k	Specified value	3,00
Wind speed, m / s	Vo	Constant	11
External diameter Rotor, m	D	Relative constant	3,00
Rotor (blades) height, M	H	Relative constant	3,00
Width of the blade	B	Specified value	1,50
Number of blades	K	Specified value	3
Area of the blade	Sn	Sn=HB	4,50
Streamlined area of wind plant, m²	So	So=HD	9,00
Mass flow of air through streamlined area, kg / sec	Mo	Mo=ρSoVo	119
Air capacity of flow, W	E	E=MoVo²/2	7187
Power take-off coefficient of flow	n	Calculation model	0,25
Equilibrium coefficient of flow deceleration	d*	d*=(4So-nc$_{fn}$KSn)/ (4So+nc$_{fn}$KSn)	0,778
Index of air energy using of wind plant	η_t	$\eta t=(1+d*)(1-d*^2)/2$	0,351
Efficiency of the wind plant	η	$\eta=\eta_g\eta_i\eta_t$	0,259
Power of wind plant,KW	N	N=Eη/1000	1,86
Specific power of wind plant, kW / m²	n_s	n_s=N/So	0,21
Loss of flow pressure in the wind engine, Pa	f	f=(1-d*²)ρVo²/2	28,63

2.7. General recommendations on technical implementation of optimal wind turbines.

The technical result of controlled winged blades using consists in essential efficiency increasing of kinetic energy conversion of airflow. Methods and devices that allow changing the configuration of the winged blade, depending on its position in the turbine and speed of rotation of the turbine.

To optimize the configuration of the winged blades of a collinear wind turbine, expediently to use a multi-segment construction (Fig.2.15).

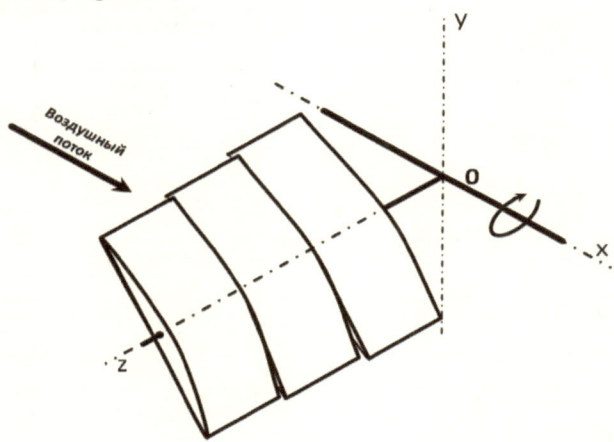

Fig.2.15. Multi-segment blade scheme
of the collinear wind turbine.

The design implies the possibility of changing the shape of the blade by orienting its segments, with using a special electromechanical drive device (Fig.2.16).

When the blade of the orthogonal turbine moves along the circular path in inversion zones the braking torque is formed against working direction of the turbine motion.

It is necessary to use the optimum regulator. On moving of a wing in these zones, the turbine switches to braking. An electromechanical trigger is switching on [65], and saltatory changing aerodynamic shape of the wing to a symmetrical one occurs. Using of flaps allows regulation of the turbine in a universal way, changing aerodynamic profile, both smooth turning of the flap and its jump-like "overturn" with using a special trigger (Fig.2.17).

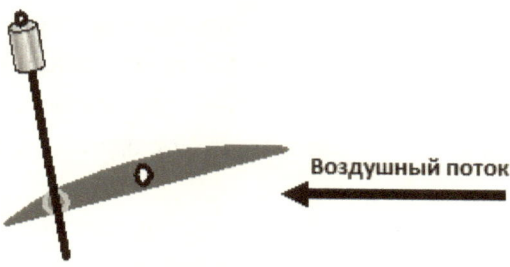

Воздушный поток

Fig.2.16. Scheme of the local orientation mechanism of a collinear wind turbine blade.

a

б

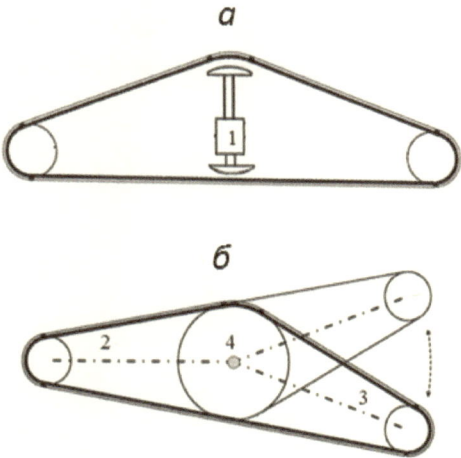

Fig. 2.17. The device of an adjustable wing blade of orthogonal wind turbines: *a*) – with trigger, *b*) with a flap; 1 - electromechanical trigger, 2 - wing, 3 - flap, 4 - flap axis.

The force moment created by the blade carries out pressure on the support axis of the blade in the working direction of the turbine rotation. It is carried out the automatic adjustment of the attack angle of each wing segment relative to the airflow to the maximum signal of pressure sensor installed on the axis of this segment.

This signal is corrected by an extreme regulator based on the controller which is located on the control node of the segment and through a special thrust rotates the segment with using a servo motor, providing the maximum local pressure on the axis of the blade.

If there is a shaft speed sensor, you can provide control of attack angles of all wings from regulators of the pressure sensor of one control wing due to the calculation of the lag of the remaining wings with respect to the control wing on the trajectory of motion in a circle with using the controller. Control commands with calculated delay are applied to flap actuators of corresponding wings. This simplifies the construction and turbine control system, increases its reliability and reduces the cost.

Summary of the second chapter

The main task of optimizing the wind engine as the main element of the power plant - maximum extraction of the kinetic energy of airflow, passing through the orthogonal section of the engine.

In the *theory of an ideal wind plant*, a wind engine is considered as a certain active zone of the airflow, in which the effect of the interaction of the flow with working engine organs are displayed in the form of flow dissipation and, respectively, reducing of its velocity (deceleration). Reducing the power of flow when it slows down is the amount of useful (extracted) power. Analysis of the equations of power and energy balance gives a calculated value of the optimal deceleration (1/3), corresponding to the dissipation of two-thirds of the stream. Wherein fraction of the extracted energy of the stream reaches a maximum, which is 16/27. This value makes sense of universal constant that does not depend on constructive and mode parameters.

In order to obtain a practical calculation technique, the theory of ideal wind plant is supplemented by private models interaction of engine blades with airflow. Basic models compiled for collinear and orthogonal wind turbines with adjustable blades of the linear-convex section, describe practically all spectrum of used wind turbines. In mathematical models, aerodynamic relationships are used and determine forces of drag and lift wing forces, the adequacy, and reliability of which is confirmed by many years of practice.

The resulting models reflect the mechanism of kinetic energy conversion of the airflow into rotational motion of the turbine, based on the assumption of superposition of the action of the applied forces, and determine the corresponding volumes of energy (power) taken out from the flow.

Calculated method formed on the basis of the model allows you to determine the optimal orientation (configuration) of turbine blades, calculate absolute and specific energy intake and power of airflow. Opportunities are realized, both accurate analytical approach and multifactorial numerical experiment.

Indicators of the optimal orientation of the blades of orthogonal turbine depend significantly on the profile of blades and speed mode of the turbine. Identified the significant effects of blade inversion of orthogonal turbines at their optimization are characterized by abrupt changes in orientation parameters (configuration).

Energy efficiency of flat-blade turbines is turned out to be essentially limited, due to a longitudinal drag of air, especially in orthogonal constructions. Thus, orthogonal turbines with flat blades are able to work only at low rotate speeds and select not more than one-fifth of the power of the wind flow. Significantly more efficient the wing blades acting on the principle of superposition of forces drag and Lift Force. They are able to function at rotate speeds many times higher than the velocity of the wind flow and provide extraction of half or more power of wind flow. A mathematical model of interaction of airflow with an impeller blade linear-convex section allows us to determine the optimal configuration of wind turbine blades, providing the maximum extraction of energy flow.

Comparison of optimization models of orthogonal and collinear turbines show a significant difference in the approaches to their optimization. At orthogonal turbine proper plane-convex, the configuration of the blade is unchanged, but its optimum orientation significantly changes with the rotation of the turbine, for which requires a special drive. In a collinear turbine, the local orientation of the blade is associated with the distance of the corresponding section of the blade from the main axis of the turbine. An optimum blade is formed not on flat and on a rule-surface base. The shape of the blade does not depend on the rotation of its reference axis and is tied only to the circumferential velocity of the turbine.

Unlike the orthogonal turbine, for which characteristic of the extreme nature of dependence energy efficiency from the peripheral speed of the turbine and respectively a drop in efficiency with high-speed modes, the use of the winged blade in a collinear turbine allows turbine for energy-efficient regimes with relatively low turnover speeds and monotonously raise the index of energy extraction of the flow with speed increasing of turbines.

The application of an ideal wind plant model together with private optimization interaction models of flow with blades of the wind plant allows getting complex methodology for the calculation of energy and power indicators of various wind plants and their comparative analysis. In particular, an analysis shows that contrary to popular opinion about limited energy efficiency indicators of orthogonal wind plants, the use of special measures to optimize them – increasing of turnover rates when adjusting the orientation of the blades practically not differing from the collinear plants with the same dimensions.

Technical realization of the above-mentioned possibilities is carried out through special constructive solutions, allowing to change the aerodynamic shape of the wing in process of its circular motion in orthogonal wind turbines. Constructions suggest the possibility of both symmetrical jumps in the shape of the wing with using an electromechanical trigger and continuous variation of the aerodynamic contour with using the rotary flap.

To optimize the collinear wind turbines it is used a multi-segment design of the blade, consisting of several local blades, located on the reference axis. The design assumes a possibility of a continuous change in the shape of the blade, by regulation of its autonomous segments with using a special electromechanical drive device.

The proposed technical solutions allow ensuring the energy efficiency of turbines by regulation of vane blades with the help of special "smart" drive, providing not only optimal orientation but also inverse change of their aerodynamic configuration.

3. Perspective developments of optimal technical solutions for wind energy plants.

For a widespread introduction of wind energy, it is necessary to lower the cost of energy in terms of both operational and capital costs. In this regard, when developing new structures and control systems for wind energy plants, first of all, optimization of their energy indicators is considered [1-3].

3.1. Collinear wind power plants with a horizontal axis of rotation.

As devices that provide opportunities for optimization, a collinear turbine of wind plants equipped with control systems of blade configuration are considered.

Wind power plant of rotating type [49] on Fig.3.1. has been analyzed together with a number of similar devices in order to develop perspective collinear wind plants.

This plant with a wind turbine 8 in the working channel 6 with confusor 1 at the input and diffuser 10 at the output, by kinematic way connected (7,13,15,17) with a rotating shaft with generator 19 and installed in the working channel in a single unit with diffuser and confusor.

On the side surface of the confusor there are automatic emergency relief valves air 2 at strong gusts of wind and on its axis 15 conical air distributor 3 with rigidly fixed on its surface guide plates 4, giving the airflow turning at a certain angle. The wind plant is located on a common rotary platform 12 having a tail plumage 11 and the rotation unit 13, which protects vertical axis 15 of swivel platform 12 from the skew.

This wind power plant is unreasonably complex design, unreliable orientation system to airflow - side areas left and right of the axis of rotation is approximately equal. Its high-speed turbine creates harmful noise and interference to electronic devices. Birds and other living creatures can fly into confusor.

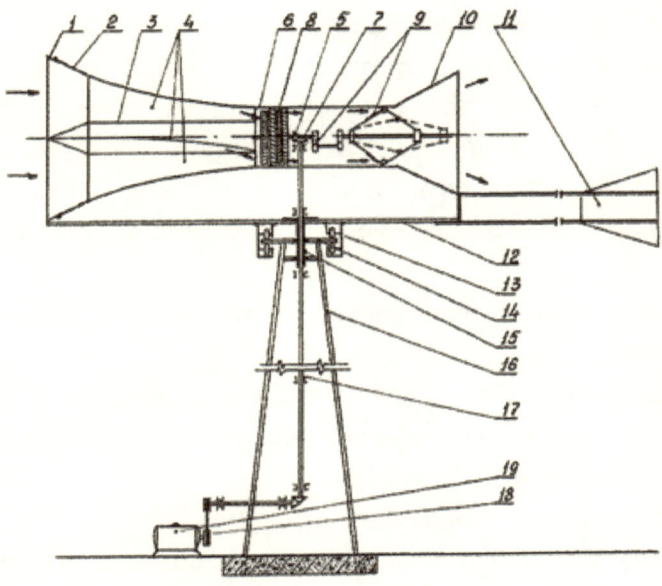

Fig.3.1. Wind power plant: 1-confuser, 2 - emergency relief valve, 3 - bailer air, 4 - guiding plates, 5 - bevel gear steam, 6 -working channel, 7 - shaft of wind turbine, 8 - wind turbine, 9-speed control device rotation of the wind turbine, 10-diffuser, 11-tail plumage, 12-swivel platform, 13 - knot rotation, 14 - fixed platform, 15 - axis rotation, 16 - support, 17- bearing, 18 - V-belt drive, 19 - power generator.

Wind plant with horizontal axis -cylindrical wind turbine [48] in Fig.3.2 and 3.3. When developing perspective devices, except basic problem of energy efficiency, the problem of creating a reliable design is solved also that meets the basic requirements for ecology - a significant reduction of vibration noise, reduce interference for electronics and harmful effects on human health, lack of external rotating parts.

The surface of the cylinder 1 is fixed to the front 17 and rear 18 wheels, consisting of rims 19, hubs 20, fixed on the horizontal shaft 2, and radial bars 21, which connect the rims of the wheels to their hubs. There are a conical reducer 5 and bushings 6 with bearings on the same shaft.

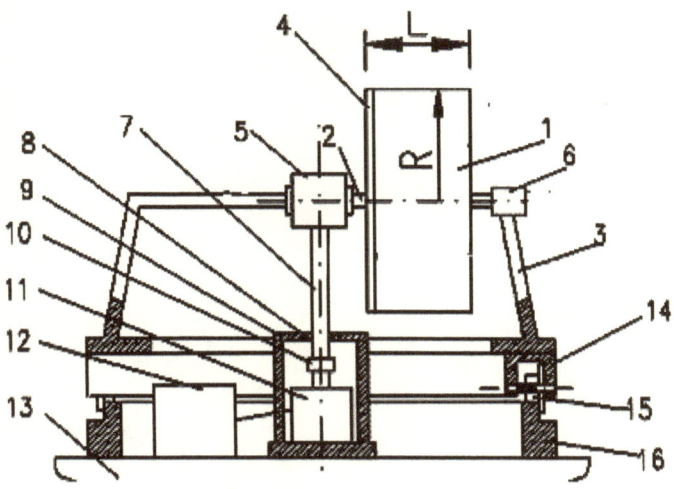

Fig.3.2. General view of the horizontal-axial wind power plant:
1 - turbine; 2 - horizontal shaft; 3 - reference design;
4- protective mesh; 5 -conical reduction gear; 6-bushing
with bearing, 7-working vertical shaft, 8 - the radial bearing;
9 - Bearing support; 10 - coupling; 11 - generator unit with
speed control; 12 – converter of electricity; 13 - roof of the
building; 14 - support wheel; 15 - wheel restraint;
16 - annular support.

Active elements - wings, have an aerodynamic profile. The wheels have N>2 rods arranged symmetrically, and the position of the rods of the rear wheel is shifted relative to the front one angle β. On the same shaft, there are a conical reducer 5 and bushings 6 with bearings. Active elements - wings 23 have an aerodynamic profile. When setting up a cylindrical wind turbine in nominal conditions it is set the angle β between the bars of the front and rear wheels, which sets the angle of attack α for all *N* wings and is fixed by the hub of the rear wheel.

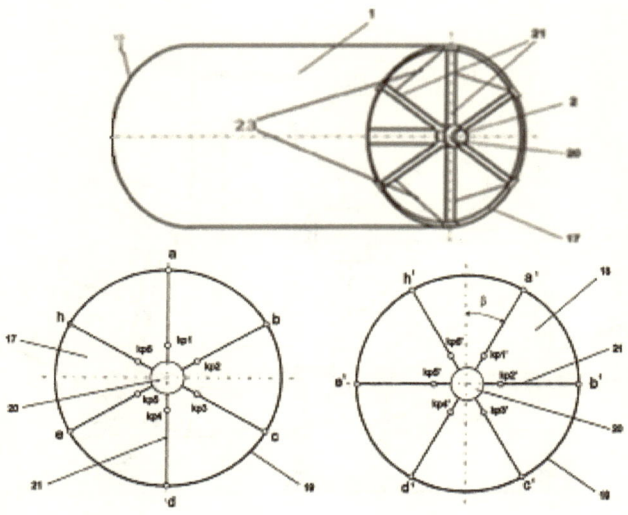

Fig.3.3. Cylindrical wind turbine: 1 - cylinder, 2 -horizontal shaft, 17 - front wheel, 18 - rear wheel, 19 - wheel rims, 20 - hub, 21 - rods A, b, c, d, e, h, 23 - wings.

Thus, a cylindrical wind turbine is a single machine with an axis, wheels, wings, ribs inside the cylinder. A feature of its construction is a permanent suboptimal position of the wing relative to the input vector of air stream - its angle of attack, set in optimal mode. The turbine is provided with an automatic orientation of a horizontal axis of the cylinder parallel to the airflow.

For this purpose, the working vertical rotation shaft of airflow orientation mechanism is placed before the front wheel of the cylinder. At any deviation of a cylinder from the direction of the airflow the force will act automatically to the outside lateral surface of the cylinder from the corresponding side and this force is proportional to the Its lateral area, which returns the cylinder to the position, parallel to the airflow.

In connection with the rigid fixation of the wings to the radial rods of the front and rear wheels of the wind turbine, it is a single mechanism with rigid fixed parts. This allows for more than at one order to reduce vibration and, accordingly, negative effects associated with them. The requirements for the strength of materials used in the construction of wings are reduced that significantly improves the reliability of wind turbines in all operating modes. Installation on the front wheel of cylindrical turbine an additional cellular mesh protects the wings from rubbish and flying living creatures. The factor that increases WEI is the effect Speed differences between the airflow, moved on the outer surface of the cylinder and the outlet airflow from the cylinder.

The advantage of the turbine is its efficient operation in a zone of low airflow rates due to a significant area of the wings. Approximate area one wing R^2, total $- NR^2$ (where R is the radius of the cylinder and simultaneously, the width of the wings), which is for an order of magnitude increases the lift force of the wings in comparison with classic horizontal-axial wind turbines.

In a cylindrical wind turbine adjustment is used mechanism for an optimal angle of attack of the wings α with respect to the airflow, this angle is selected when tested in nominal mode: nominal air velocity, nominal cylinder revolutions, and rated load.

The developed wind turbine, thanks to environmental benefits and design reliability, can be used in a residential area in the locality with low average airflow rates. A disadvantage of the wind turbine cylindrical design is that the design of a cylindrical wind turbine allows you to change the angle of attack α in a very small range, several degrees. In this way optimization of the angle of attack of the wings, α is carried out only for some nominal operating mode, reducing turbine efficiency.

Prismatic wind turbine [50] is a modification of cylindrical wind turbine, which is characterized by a replacement of cylinder 1 in Fig.3.2 and 3.3 on the prism 1 of Fig.3.4, with adding special control devices.

The turbine retains the advantages of a cylindrical wind turbine but eliminates its above-mentioned shortcomings.

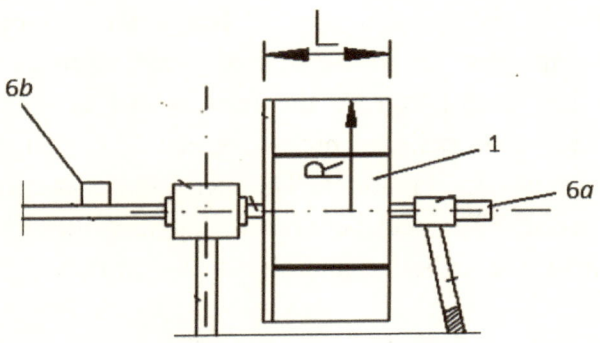

Fig.3.4. Prismatic wind turbine – node of wind turbine:
1 - prism; *6a* - control device of a turbine; *6b* - airflow velocity sensor.

The purpose of developing a prismatic wind turbine is the creation of a simple and reliable design, which meets the basic requirements for ecology and providing the maximum possible energy extraction of airflow. This is achieved by choosing the optimal angle of attack α of all wings, at which maximum possible rotation speed of the main horizontal shaft of a prismatic wind turbine is guaranteed with a specific current air velocity of flow and load on the generator (on the working shaft of the turbine).

Aerodynamic wings 21 of length L and width t are active elements of a prismatic wind turbine (Fig.3.5).

The basis of the prismatic wind turbine is an m-sided prism, which is formed by means of symmetrical polygons 17, 18 in the front and back bases (Fig.3.6).

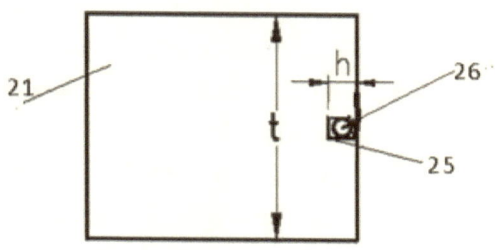

Fig.3.5. Wing, top view: special groove -25, nut -26

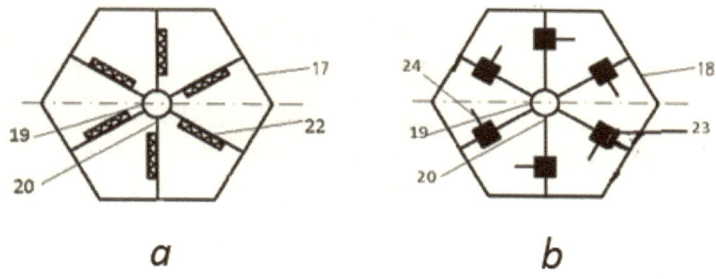

a   b

Fig.3.6. The bases of the prism: the front one -17, the rear one - 18, hub -19, rod -20, hinge -22, WAAA -23, axis with thread-24.

Hubs 19 fixed to the front and rear the ends of the main horizontal shaft 2 are connected by means of the rods 20 with the corresponding midpoints of the corresponding polygon sides 17, 18 – bases of a prismatic wind turbine at an angle of $\pi/2$. Prismatic wind turbine contains aerodynamic wings attached to the rods 20. A quantity of sets of front and rear bars is m. Front edges of the aerodynamic wings 21 are fixed to the front set with the help of loops 22.

In the middle of the rear set of rods of wing attack angle adjusters (WAAA) 23 are fixed – tracking electric drive with sensor and position controller, which sets the angle of attack α to all aerodynamic wings of the control unit 6a with extreme regulator of rotation of the main shaft at the base of controller receiving information from the rotation shaft sensor in the control unit. WAAA 23 rotates the threaded axles 24, in which in special grooves 25, the nuts 26 are pushed with the possibility of moving in limits h with a change in the angle of attack α and fixed to the rear edges of the aerodynamic wings with the rear edges of all the aerodynamic wings which move synchronously under the influence of WAAA 23 by commands from the node control (from the extreme regulator).

A feature of the prismatic design of wind turbine is to maintain the optimum angle attacks of all aerodynamic wings relative to the vector of airflow in operating mode. In this mode due to a selected angle of attack α front edge position of each aerodynamic wing will be different from its trailing edge relative to the airflow vector approximately by the amount of H - the difference between the levels of the anterior and the trailing edge of the wings (see Fig.3.5).

$$H = L\ tg\alpha \tag{3.1}$$

where L is the length of the prism (in the example $L = R$).

When testing the angle of attack α in Fig.3.7, the rear edge of wing 21 is shifted towards the turbine inlet by an amount

$$h = L\ tg^2\alpha. \tag{3.2}$$

To protect the turbine in the control unit 6a the monitored parameters are previously entered: limit values of air velocity V_n, revolutions of the main horizontal shaft – n_n, the limiting current value of the generator – I_n, and when the signals received from the respective sensors, exceed limiting values of controlled parameters, the controller switches inputs of WAAA from the output of extreme regulator of revolutions on inputs of regulators tracking the relevant parameters on the base of controller, which through the WAAA deflect the attack angle of wings from the optimal value of α and prevent exceeding of the monitored parameters, that is, reduce the angle of attack α of the wings down to zero.

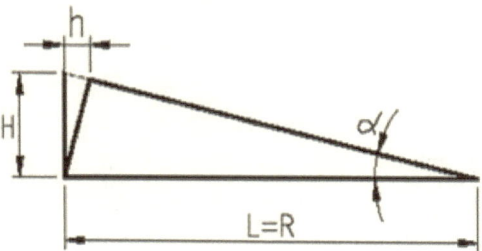

Fig.3.7. The change in position of the aerodynamic wing in process of adjusting its angle of attack: h-displacement of the rear edges wing.

A prismatic wind turbine can be used in residential complexes with unreliable power supply, and also in cases of natural disasters as a reserve source, for which the airflow speeds at 10-15 m/s are completely working. Recommended power of turbine is of 10-40 kW.

The centralized orientation method of horizontally-collinear wind stations. Large wind stations are actual projects of scale power supply, since the solutions, aimed at the development of renewable sources of energy, are now a priority for the majority of developed countries. According to some estimates in countries where wind energy is actively developed, it already can compete with thermal energy sources with a cost of energy per 1 KW-Hour. However, the current airflow orientation systems complicate design of the unit and reduce its reliability.

Among known horizontal-axial wind plants, the closest in technical essence to the developed technical solution is wind plants produced by Anhui Hummer Dynamo Co. Ltd [51]. By the method realized in this wind plant a horizontal-axial turbine is placed with one side of the vertical axis of rotation and the weathervane made in the form of vertical tail is placed on the other side of this axis. In this case, the product of the weathervane area and distance from the vertical axis to the geometric center of the weathervane is chosen so that it exceeded the product of the area of the lateral projection of the wind wheel at a distance from the geometric center of this projection to the vertical axis. In this way center of lateral pressure of the entire horizontal structure, rotating about a vertical axis is located behind (relatively to the direction of the airflow) of this axis. The construction becomes statically stable in accordance with relative to the airflow and for any change of flow direction, the wind wheel plane of rotation will always turn into a position perpendicular to the direction of flow.

Constantly increasing the value of land suitable for the placement of large power wind farms leads to increased problems with their accommodation. Fig.3.8 shows realization option of the proposed method [52].

Platform 1 is placed on the surface of the pond (sea, lake, reservoir, etc.). In the frontal part of the platform, in its diametric plane which is vertical to a plane of symmetry, one end of flexible traction is hinged (for example, anchor chain 2a), the other end of which is hinged at anchor 2.

The center of the lateral pressure of the whole structure usually provides static stability (position of the axes of all plants is parallel to the airflow). Each wind plant 4 is fixed to the top of the corresponding tower 5. However, in some constructions platform area is not enough to stabilize the position relative to airflow (for example, a platform based on pontoons).

To increase the static stability of the platform with several wind plants mast 3a is installed on its stem with a sail 3 (plane made of plastic, ceramics, sheet metal, etc.), which fixed in the diametric plane. As a result torque of force will be attached to the construction of platform relative to a vertical axis passing through the anchor 2 (with a large shoulder relatively to the mast of the sail 3a). This torque of force will deploy the platform to a position where the axes of wind plants will be parallel to the direction of the airflow.

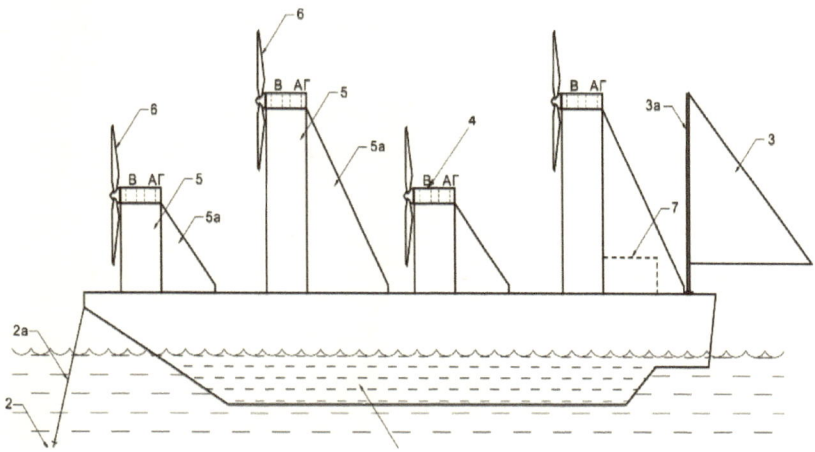

Fig.3.8. General view of the floating wind farm: 1 - platform; 2, 2*a* - anchor and anchor chain; 3, 3*a* - mast with feed sail; 4 - wind plants, variable speed driver, generator; 5, 5*a* - the tower and its support; 6 – wind plant propellers; 7 – power block.

For a stable orientation of the axes of wind plants such platforms are used which design Is symmetric to their diametric plane. In some cases on exceeding of the nominal airspeed it is advisable to limit the speed of wind turbines by diverting the plane of the sail 3 from the plane at some angle β manually or automatically, for example by means of information from the flow velocity sensor when the appropriate electric drive for the stern mast 3*a* is available.

The proposed technical solution for the group of wind plants of considerable capacity allows eliminating negative environmental factors and place plants in areas with minimal wind resistance. In addition, in this case, the design of wind plants is simplified, their reliability is increased, the cost is significantly reduced. In connection with the removal from the coast, it is advisable to accumulate the generated electrical energy in batteries located, for example, in universal energy containers. This simplifies delivery and connecting containers with uncharged accumulators, removal from the platform containers with already previously charged batteries. A variant of charging circuit of batteries has been developed (see Fig.3.9).

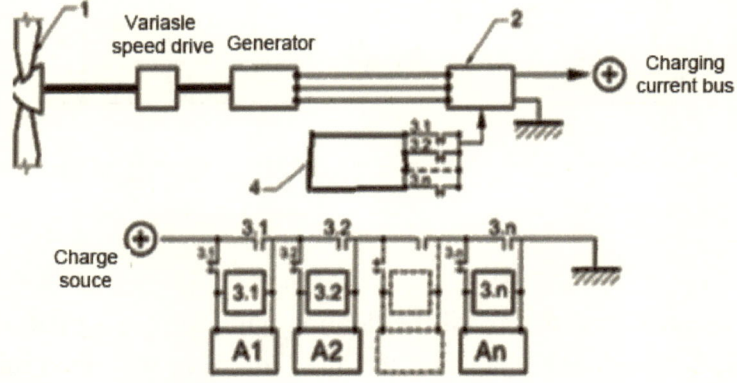

Fig.3.9. The charging circuit of the batteries:
1-propeller, 2-current regulator, 3-1, 3-2 ... -voltage relay
and their contacts, 4-current controller, A1, A2 - batteries.

The proposed method allows you to reduce the cost of wind stations by at least 30% and simplify their construction without the cost of scarce land on the shore.

A number of advantages of the proposed method are noted. Instead of the vertical axis around which in classical case gondola of the wind orientation to the direction of the airflow is rotated, in proposed method virtual vertical axis passing through the point of the hinged anchor fastening is used. In this case, you are allowed to orient all wind plants in centralize way without orientation mechanisms of the classical construction (watching system, gondola with a turn ring).

The rotational transformer i (or Ring current collector) for transmission received electrical energy is eliminated.

In the proposed method for orienting several, the weathervane function is performed by the housing of platform with a stern sail.

To place the wind farm, you can use areas with minimum wind resistance, not demanding land plots with unacceptable ecology.

3.2. Collinear wind power plants with the vertical axis of rotation.

The use of vertically-collinear structures allows combining energy efficiency of high-speed horizontal-axle wind plants with advantages of vertical-axial ones – greater dynamic stability, the simplicity of design, lower production and maintenance costs of wind stations.

Vertical wind power station with a concentrator of airflow [53] (Fig.3.10). An increase of wind power station efficiency is provided by the fact that the station contains an electric generator connected by the vertical shaft with horizontal wind wheel paced inside a pipe, conjugated with a concentrator of airflow.

The airflow concentrator consists of lower and upper pyramids 1 and 2, coaxially located by their tops against each other. Pyramids 1 and 2 along their edges are interconnected by trapezoidal walls 3, as a result channels have been formed between them constricted from periphery to the center. The edges of the lower pyramid 1 are united by a plate 4, in the center of which is installed electric generator 5 connected by a vertical shaft 6 with horizontal wind wheel 7 located in the pipe 8.

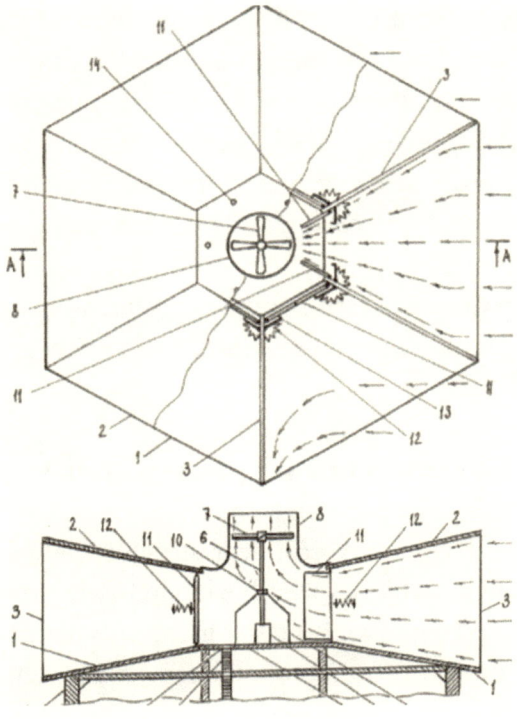

Fig.3.10. The design of a vertical wind power plant
with airflow concentrator.

The lower part of the tube 8 is expanded and interfaced with the upper pyramid 2. Under the influence of wind of any direction, for example, as indicated by the arrows in Fig.3.10, the flow enters the concentrator channel facing the wind, and because of its narrowing this flow contracts and increases its speed. Then the airflow passes opened by its action door, closes with his pressure other doors 11, pressing them against spring stops, and rushes to the wind wheel 7, giving it a rotary movement, which is transmitted through the shaft 6 to the generator 5.

When the direction of the wind changes, the airflow enters into other one or two adjacent channels and also, accelerating its movement, is directed to the wind wheel. In case of a hurricane wind, the airflow passing the channel directed against it opens the doors of other channels by its pressure turning the slats 13 and compressing the springs 12, which are designed for the beginning of their compression at the maximum wind speed, safe for the wind wheel 7. The force acting on the door at this wind speed is determined by the aerodynamic calculation. Through the open doors 11 part of the airflow breaks out, weakening its strength in the direction of wind wheel 7.

The higher the speed of a hurricane, the doors 11 are wider open, compressing the springs 12, and, accordingly, most of it is pulled out. In this situation wind wheel 7 does not experience excessive stresses that exclude its damage. After finishing action of the hurricane wind doors 11 by forces of springs 12 are returned to the original (closed) position, and airflow acts on the wind wheel 7 with the permissible force.

Excess of energy generated by the generator 5 during the reduction of its consumption is transmitted for charging of batteries that are used for supplying energy to consumers during periods of calm. The power of this electric power plant is much higher than power of a conventional electric power plant that does not have airflow concentrator, with the same dimensions of wind wheel. This is due to the fact that in the usual electric power station wind power is used only with vertical area equal to the area of the wind wheel, and in this power plant wind power is used from vertical area of the outer part of the canal concentrator, which is several times larger than the area of wind wheel.

So, for example, with the ratio of the vertical area channel to the area of the wind wheel, which is equal 10, and taking into account the losses to overcome the airflow of frictional forces by 30%, the wind wheel receives approximately 7 times more energy than when it is open. At the same amount power plant capacity is increased. Depending on the size of wind wheel and airflow concentrator, a station can reach a power of 5-10 megawatts.

Vertical-collinear wind energy towers are perspective structures in form of vertical cylindrical, conical and other shells containing collinear wind turbines, and give significant scalability of wind plants. As an example, in Fig.3.11 the design of the inverse energy tower is shown.

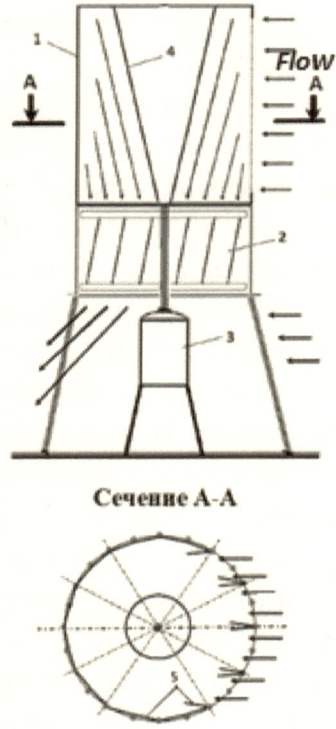

Сечение A-A

Fig.3.11. Inverted collinear wind tower:
1 - wind receiver; 2 - collinear wind turbine;
3 - generator; 4 - the guide cone; 5 – flaps of wind receiver.

The peculiarity of the design is that airflow trap is a wind sump 1, is located at the top of the tower where energy potential of the wind is higher and the power unit includes collinear turbine 2 and electric generator 3 - at the base of the tower. Such arrangement allows providing comparative simplicity of the wind engine design, dynamic sustainability, and accessibility for maintenance of the power unit. Like the previously discussed construction the wind receiver is equipped with special flaps which are opened under the pressure of wind (with a windward side), and blocking the airflow with the opposite side which eliminates the need for orientation of the wind power plant in wind direction.

3.3. Orthogonal wind energy plants with unregulated and self-regulating working bodies.

The considered plants represent attempts of relatively simple means, without using automatic control systems, to optimize devices for the extraction and canalization of an energy of airflow.

The sailing wind power plant [54] in Fig.3.12, capable to convert the kinetic energy of airflow in mechanical reciprocating motion, is considered as a prototype of the device. It directly converts the frontal effects of airflow in various types of energy.

At the same time problems of sail orientation relatively to flow direction, transferring to weathervane mode (with the return of the mast) and again the installation of active sail. To implement the device, you need to solve problems of control for cyclic operation of the sail. The disadvantage of this wind power plant is a low efficiency, due to the fact that the mast on the return course does not generate energy (dead time).

It is known that the efficiency of the compensators is 0.5, the efficiency of hydraulic converter – 0.75. If maximum WEI of sailing plants − 0.197, then in this installation effective WEI is much less − 0.197 x 0.5 x 0.75 = 0.074.

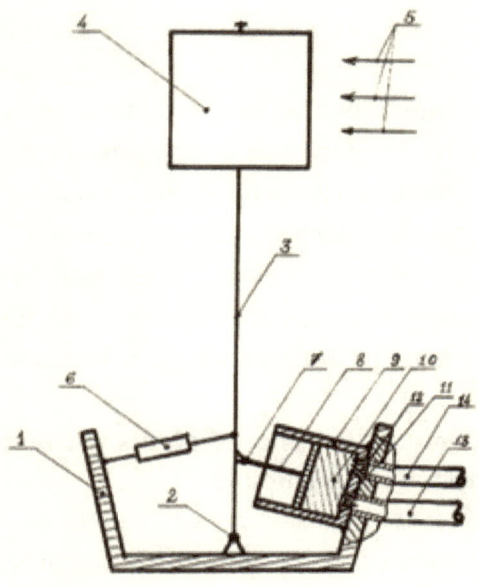

Fig.3.12. Wind power plant: 1 - support,
2, 7 - hinges, 3- mast, 4-sail, 5- flow, 6-compensator, 8 - piston,
9 - pump, 10 -working chamber, 11, 12-inlet and exhaust valves,
13, 14-supply and pressure tube.

Wind power plant - wind pump [55] - improved plant according to Fig.3.12. A technical result (Fig.3.13) is to ensure its reliable cyclic operation regardless of the direction of the airflow and variations in its velocity. The plant contains a support, a rotating shaft, which is fixed in the upper and lower support cups with bearings.

The lower support cup is connected to the surface of a support, and the upper one is connected to the structural unit, providing a vertical position of the shaft attached by one end to the support and the other one is connected to a device that has surface area and the ability to resist flow - a rotary carousel wind wheel.

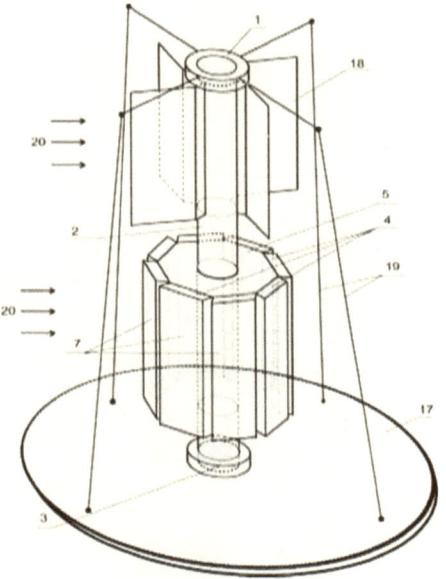

Fig.3.13. Pump for wind: upper supporting cup-1, shaft-2, lower support glass-3, power unit-4, rectangular prism-5, membrane-7, fixed support-17, wind-wheel-18, structural unit-19, airflow-20.

Below the shaft, there is symmetrically mounted rectangular multilateral prism, to each lateral plane of which is attached power unit, made in the form of the membrane and elementary pump with a piston that has inlet and outlet valves connected to the supply and pressure mains. The plant can be used for work with electric motor having a screw through converter connected to the batteries, previously charged, for example, at night during parking.

Such plant (see Fig.3.14) can be used in agriculture. It can be used as a pump for pumping liquids and gases, lifting the liquid to a height (for example, for a water tower), injection of air under pressure in receivers and further use in pneumatic tools and mechanisms.

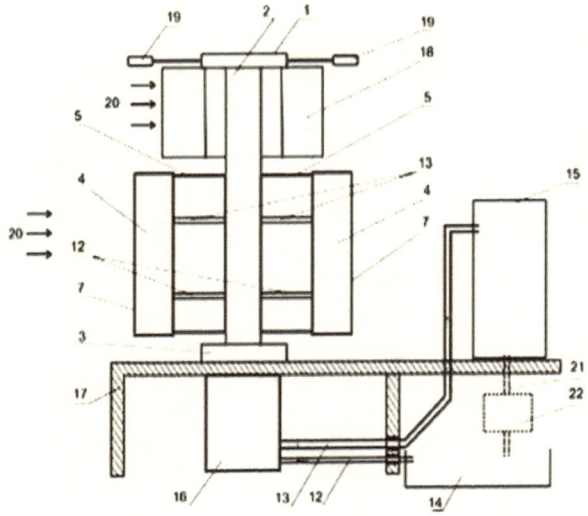

Fig. 3.14. Pump for wind for agriculture: top support cup-1, shaft-2, lower support cup-3, power node-4, rectangular prism-5, rod-6, membrane-7, piston-8, pump-9, pressure - 10 and suction - 11 valves, supply-12 and pressure-13 mains, rotary joint-16, fixed bearing-17, wind-wheel -18, structural node-19, wind flow -20

Carousel wind wheel [56] on Fig.3.15. The design has been obtained as a result of improvements of the famous carousel wind wheel [26] on Fig.1.18 (the upper tier without the upper separation disk and a side view for the option of a two-tier wind wheel).

To increase the power and reliability of the structure, each set of L-shaped flaps is sandwiched between two dividing planes (disks), structurally forming one tier, sealing the airflow, it is directed to the blades since the separation planes prevent flow sliding down from the top and bottom side of the blades. Relation α = 2π/kn provides symmetry of construction and rotation uniformity of the considered wind wheel. Axis of rotation 3 of each blade 2 coincides with its nearest lateral edge. In the operating mode, the blades 2 are pressed against the thrust frame 4 fixed parallel to the plane of each blade on the lever 1, to its construction through dampers 5.

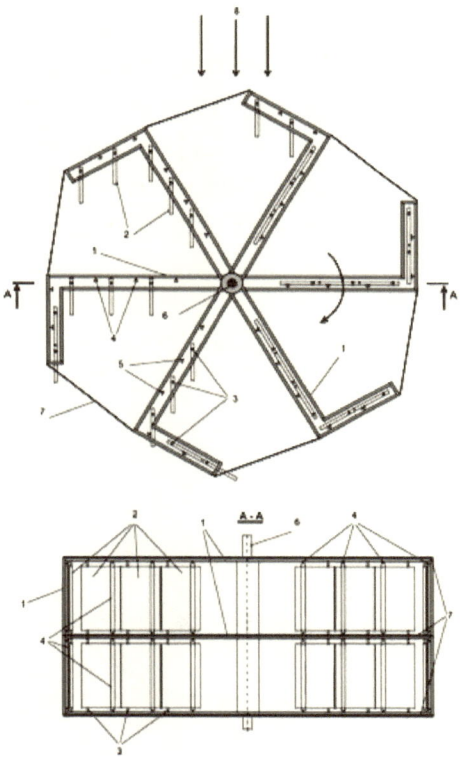

Fig.3.15. Carousel wind wheel: 1 - L-shaped swings; 2 - blades; 3 - axis of rotation; 4 - stop frame; 5 - damper; 6 - vertical axis of the wind wheel; 7 - disks; 8 - airflow.

Like the well-known wind wheel [26], it works in 3 quadrants relative to the airflow vector, and a group of blades on the radial direction work in the 1-st and 2-th quadrants and a group of blades located on disks perpendicular to the radial operates in the 2-nd and 3-rd quadrants. At a flow rate exceeding permissible values, parallel movement of the thrust frame 4 is provided which is less than the width of blade 2 that leads to the appearance of gaps between blades and power take-off reducing.

In the weathervane mode, the blades 2 as far as they move rotate in the range from 0 to 180 degrees in the direction of flow and do not have in this range of angles any restrictions. A clear defect of both wind wheels is an overpriced number of blades.

Carrousel wind turbine - turbine [57] in Fig.3.16. With the same efficiency of work as in [52] - works in 3 quadrants relatively to the airflow vector, but a number of blades reduced 2 times. This simplifies the design of the wind wheel, increases its reliability.

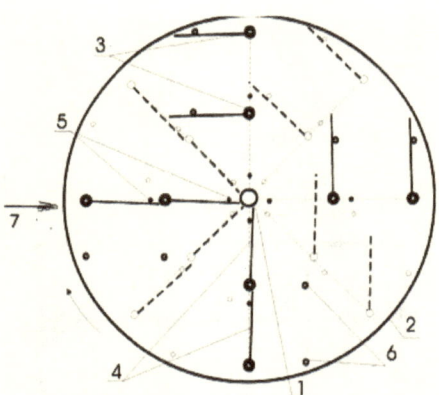

Fig.3.16. Details of the turbine, top view: 1-shaft, 2-separating discs, 3-axis hinges, 4-flexible blades, 5, 6-delimiters, 7-flow work mode.

Separating discs also concentrate the flow on blades 4, do not allow its disruption along the upper and lower Edges of blades. Under the action of the flow 7 of the blade 4 of each radial direction (see Fig.3.16) are rotated around the hinge axes 3 between the stops 5 and 6 depending on the angle between radial direction of the disk 2 with the blades 4 and the direction of the flow. At a flow rate, not exceeding the nominal, the blades of each t turbine tier operate in two modes.

Working mode. When the blades 4 are in the 1-st - the 3-rd quadrants (0 - 270 degrees) between the current position of the radial direction of the blades 4 and direction of the flow 7 of the working fluid, the blades 4 are pressed against limits 5 or 6. The force from the pressure of the working fluid on blades 4 through hinges 3 and stops 5 or 6, and also through the disk 2 is transmitted to the shaft 1 of the turbine, on which the moment of rotation is realized. In this case, overflow working fluid from the zones of increased pressure on the blades 4 turbine to the zones with reduced pressure occurs only through the edges of those blades 4 (with their radial orientation), which are proximal to the edge of disk 2 and also through the edges of the blades 4 located normally to disks 2, with different orientation of the blades 4 (in the 3rd quadrant). At the same time, the flow of the working fluid through edges of the blades 4 parallel to the disk 2, between the edges of neighboring blades 4 (with their radial orientation) and also between the edges of the blades 4 proximal to the shaft 1, and shaft 1 is practically absent.

Weathervane mode. Implemented when blades are placed 4 in the 4th quadrant of disk 2 (relatively to the vector flow). In this mode, the blades 4 are parallel to the flow 7 of working fluid, pressure on both sides is the same and, as result, the moment from these blades on shaft 1 in this mode is not created. When the load on the blade 4 exceeds nominal one, for example, at a storm speed of airflow, each stage of the turbine with blades 4 works in the overload mode. Due to the elastic properties of blades 4 and stops 5 and 6 between the edges of adjacent blades 4 there are gaps through which the flow

Flows from the area of increased pressure on the blades 4 in a low-pressure area. The pressure difference between opposite surfaces of the blades 4 are reduced, at the same time moment of force on shaft 1 is decreased and as a result, the rotational speed of the carousel wind wheels reduced to a value that does not exceed permissible one.

Due to the fact that the blades considered above wind wheels operate only in three quadrants the task has been put: to develop a wind wheel that operates In 4 quadrants, that is, fully using the energy of flow within its streamlined area. In Fig.3.17 it is shown carousel wind wheel without top spacer disk in which the flat blades 1 are located on the rods 3 with equal angles between them.

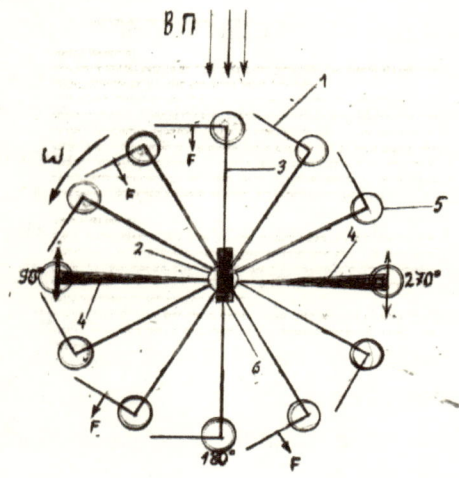

Fig. 3.17. Carousel wind wheel: 1 - blade,
2-shaft wind wheel, 3-rod, 4-zone of switching,
5-electric drive with a position sensor, 6-control unit.

In order to provide a given direction of rotation of the wind wheel, its blade 1 in the first and fourth quadrants (0-90, 270-360 degrees) relative to the vector of airflow, should be perpendicular to their rods 3 and are to the left of them relatively to the shaft 2 and to the right of their bars 3 relatively to the shaft 2 in the second and third quadrants (90-180, 180-270 degrees). On the drawing, it is also shown effect of the forces F from the flow on various sections of blades motion trajectory1 creating torque in the selected direction.

In switching areas 4 (Δ degrees) it is necessary for preservation of the direction of rotation (sign of the moment) to turn the blades 1 at 180 degrees from the transiting from zones with the left position of the blades 1 on the rods 3 relatively to the shaft 2 to the position of the blades with the right arrangement of the blades 1 on the rods 3 relatively to the shaft 2. To implement this algorithm for controlling the wind wheel electric actuator 5 is required with a position sensor on the axis of each blade and the control unit of the wind wheel 6, which includes weather vane 6.1, shaft position sensor 6.2, stator of which is rigidly connected to the weather vane 6.1, and the controller 6.3, receiving information from 6.1 and 6.2 for management of electric drives 5 of blades 1. At the same time, it is shown a simple technical solution to the problem in Fig.3.18 without electric drives, weather vane, shaft position sensors and also the controller.

The wind wheel is equipped with a pair of blades 3 on each of the N T-bars 2. Blades 3 are pivoted to axes 2, and the axis of rotation of each blade coincides.

The first of each pair of blades is pressed against limit-damper in the first zone - to the left of the bar relative to the shaft 5 (0-90 degrees, 270 -360 degree relative to the airflow vector) and creates torque in the selected direction. The second blade - in the second zone, to the right of the rod relative to shaft 5 (90 degrees-270 degrees relative to the vector of flow), is pressed against the limiters-dampers and creates torque in the selected direction (at this moment the first blades operate in a weathervane mode). Then when moving to the first zone, the first of each pair blades is pressed against the limiters-dampers 4 and creates torque in the chosen direction, and the second blades at this time, transfer to a weathervane mode.

The considered version of the wind wheel provides effective extraction of energy from a unit of streamlined area, greatly simplifies the design, improves reliability and reduces capital investments. Fig.3.18 shows that each set of blades 3 and dampers 4 on each T- rod 1 at any point in the circular path of motion has blades 3 in operating mode (with an emphasis on limiters - dampers 4 that creates a moment and blade 3 is in the weathering mode, at the same time the rotational moment of each set is not interrupted, passing from the first zone into the second one, and so on.

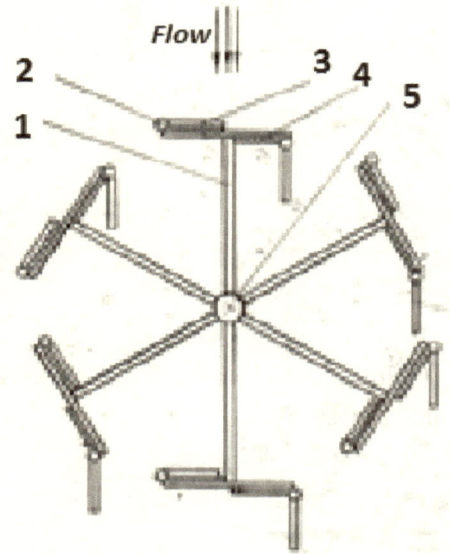

Fig.3.18. Wind-wheel: 1 - T-shaped rod, 2 - axes of blades, 3 - blades, 4-limiters-dampers, 5-shaft.

Carousel wind wheel according to [58] provides work In 4 quadrants of airflow, that is, uses the energy of the flow within the entire streamlined area. At this wheel idea of the variant of the wind wheel according to Fig.3.16 are used.

The proposed wind wheel is shown in Fig.3.19-3.21. Flat blades 6 are fixed in the form of several disjoint equal chords on the upper 3 and lower 4 discs rigidly fixed to the main shaft 1 and used as dividing planes.

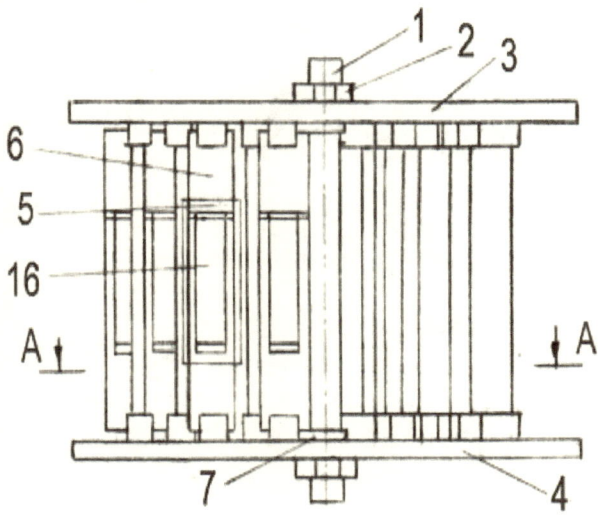

Fig.3.19. General view of the wind wheel: 1-main shaft, 2-nuts, 3-Upper Disc, 4-Lower Disc, 5-knot inserts, 6-blades, 7-support rings, 16-insertion

Radiuses from the main shaft 1 through the centers of these chords form between themselves equal angles $\alpha = \pi/k$, where k - a number of chords on disks.

The main property of such wind wheel is uniformity rotation improvement since its blades are created continuous torque in 4 quadrants. During high speed of airflow rotation speed of the wind wheel is limited by opening the inserts 16 to blades.

Fig.3.20 shows the details of the wind wheel for a variant of 3 chords and directions of flow and rotation of wind wheel with arrow ω are given.

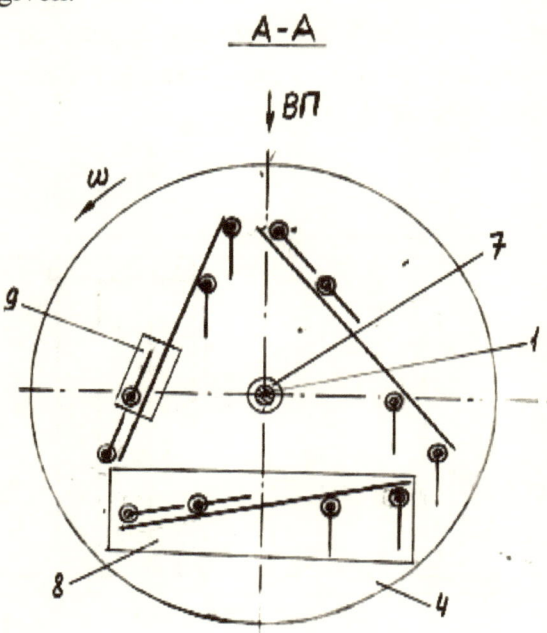

Fig.3.20. Section of a wind wheel on AA: 1-main shaft, 4-lower disc, 6-blade, 7-stop ring, 8- set blades and dampers on the chord, 9 - axis with damper.

Fig.3.21 shows the design of the insertion site 5. Each set of blades and dampers 8, on each chord in any point of the circular motion trajectory of the wind wheel has both blades 6 in operating mode, which creates a rotating moment and blades 6 in the weathering mode, that is, the torque of each set is not interrupted.

This circumstance provides uniformity of rotation of the proposed wind wheel and eliminates the need for several tiers. The insert 16 opens slightly at high speeds of airflow, reduces the streamlined area and the actual torque of the wind wheel, significantly widening the working range of flow rates without exceeding the limiting speed of rotation.

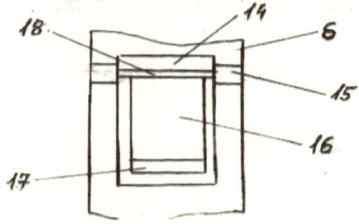

Fig.3.21. Insertion site: 6-blade, 15-axis support
of the insert, 14 - axis of insertion, 16-insertion,
17-load bar,18-inset window.

Fixed meshed net on the edges of the upper and lower discs protects the construction from birds and debris.

3.4. Optimal control systems of wind turbines.

Optimization of wind power plants proposes an implementation of regulation concerning some directions [59-63]: orientation of the wind engine along wind direction, configuration (orientation) of working bodies of the wind engine, protecting of the plant during extreme modes of operation, regulation of extraction of energy.

Application of wind control mechanisms for positioning the wings {blades}in the design of vertical-axial plants allows you to rotate them relative to the traverse so that the magnitude and action direction of the resulting aerodynamic force on the wings allow self-starting, even at airflow speeds of 3 to 4 m/s.

In addition, the control of the wings position gives ability significantly to improve the performance of such wind plant. With active control of wing's angles of their rotation relative to the traverse, taking into account direction of flow at each point of the circular path, are implemented by a special mechanism.

The method of transforming the kinetic energy fluid in the rotational motion of the wing and a plant for carrying out this method [64] (see Fig.3.22).

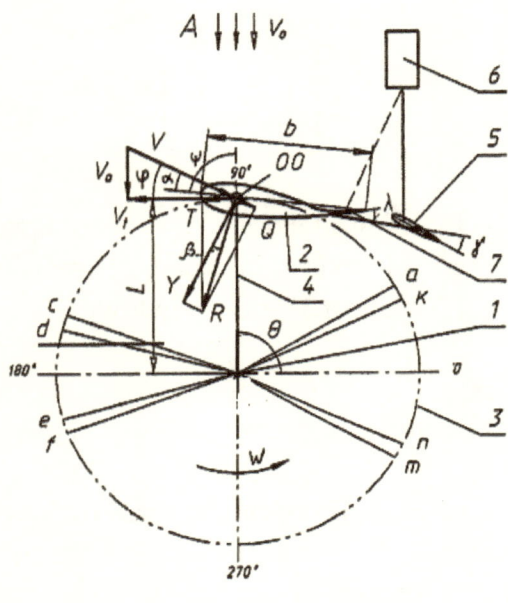

Fig.3.22. Wind turbine with adjustable wing: 1 - axis, 2 - wing, 3-circular orbit, 4-rod, 5-winged Element, OO-own longitudinal axis, 6-drive management.

The technical result is that axis 1 is set perpendicular to direction of flow A of this fluid and at a certain distance from this axis the wing 2 is placed which own longitudinal axis OO is parallel to the axis 1, around which the wing 2 under the action of hydrodynamic forces R acting on it, commits rotational motion along a circular orbit 3 and oscillatory motion around its own longitudinal axis OO and when the wing 2 moves along an arc of a circular orbit facing the flow A of the fluid.

On the arc (a,c), the angle α of the wing attack 2 is constantly kept with one sign. And on moving wing 2 along the opposite arc of the circular orbit (f,m) sign of the constant value of the angle α of the attack of the wing 2 is changed also on the opposite, and on an arc circular orbit, on which the wing 2 moves against the direction of flow of the fluid flow A (k,n) and the circular arc orbit, on which the wing 2 moves in the direction of motion of this stream A - d, e the value of the attack angle is set equal to zero.

Angular position actuator of wing-like element 5 relatively to the wing 2 can contain a cam mechanism (Fig.3.23), a cam 9 which is fixed on the fixed axis 1.

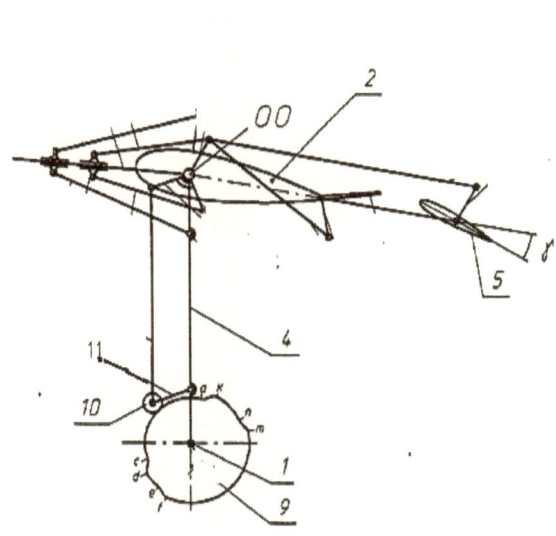

Fig.3.23. Paddle drive actuator: 1-axis, 2-wing, 4-rod, 5-winged element, 9-cam, 10-roller, 11-rocker.

The profile of the cam 9 is formed by four arcs. Two the "de" and "nk" arcs of the same radius are located diametrically opposite, one arc "ac" has smaller, and the other "fm" larger radii. All arcs are smooth transform one into another. The cam 9 interacts with the roller 10 of the rocker 11, which is mounted on the rod 4. The roller 10 is pressed to the cam 9, for example, by springs 1.

The disadvantages of this design include:

a necessity to orient cam 9 to Direction of the airflow, for example, through the weather vane rod with selsyn-sensor and selsyn-receiver setting the cam position;

a quite complex structure for control of each wing for implementation;

a divisible decrease in angular momentum at a negative angle of attack α, equal to positive, leading to a decrease of the effectiveness.

Wind power plant with vertical axis [65]. The general view of the plant is given in Fig.3.24.

The most obvious principle of optimization plant operation is chosen - maximum force, it creates the moment of rotation specified by each of its wings, that is, in fact, with a certain precision, dependent on the control system accuracy and structures of some of its nodes, optimization criterion for maximum moment for each wing.

The technical result of the proposed solution Is a significant increase in efficiency of kinetic energy conversion of the airflow. Under the influence of the flow on each wing of the installation, with moving it in a circular orbit, a permanent torque is formed, tending to maximize possible, at each point of the trajectory for a given speed of airflow. The axis of each wing 7 passes through a positioning node of wing axis (PNWA), in Fig.3.25. The wing itself is three-layered. It is based on the base sheets, to which along with the front and back edges two-sided mobile figured aerodynamic sheathing 17, 18 is fastened by means of hinges (Fig.3.26).

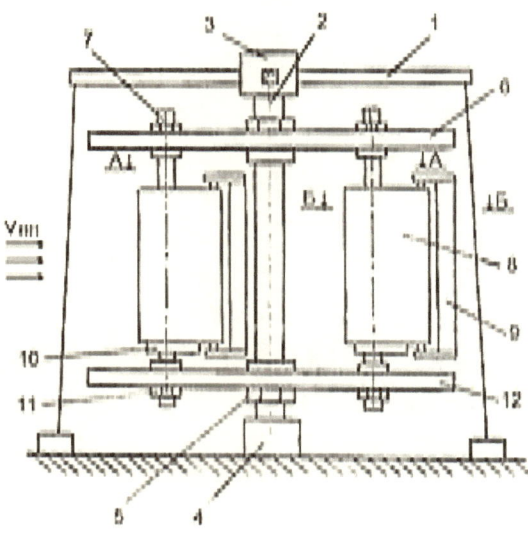

Fig.3.24. General view of the wind plant: 1 - frame,
2 - shaft, 3, 4 - shaft supports, 7 - wing axis, 8 - wing,
9 - flap, 5 - nuts of a shaft, 6, 12 - disks.

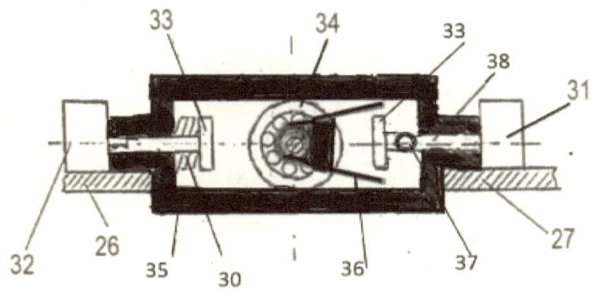

Fig.3.25. Unit of wing axis positioning: 7- wing axis,
30-spring, 31-sensor, 33 -pusher, 34-bearing,
35-body PNWA, 36-catcher, 37-hinge.

The aerodynamic shape of the wing is changed abruptly,
symmetrically and relatively to the base sheet 26, 27 at a

movement of the wing along a circular orbit in the switching zone 13 (changes in the sign of the moment formed by the wing, the Fig.3.27).

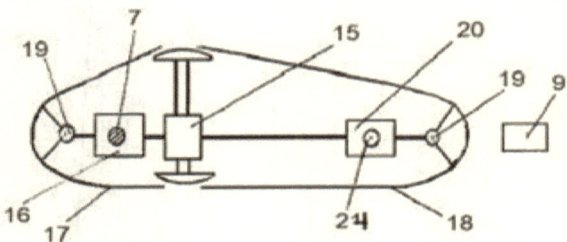

Fig.3.26. The design of the wing: 7 - axis of the wings, 15-trigger, 16 - PNWA, 17 - front JSC, 18 - rear JSC, 8 - wings, 9 - flap, 19 - AO axes, 20 - control unit, 24 – axis of the flap.

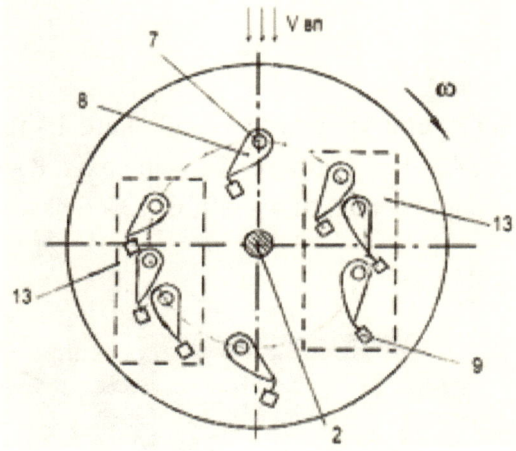

Fig.3.27. The trajectory of the wing movement: 2-shaft, 7- wing axis, 8-wing, 13 - zone of torque change.

This signal adjusts the extreme regulator to the base of the controller which is located in the control node (pos.20, Fig.3.26) and through the axis of the flap 24 rotates flap with servo motor, providing a pressure of wing on its axis.

When the wing moves along a circular trajectory in zones 180 degrees + and 360 degrees +, relatively to the resultant airflow, in zone 13 of the change of the wing shape, the braking moment is formed counteracting the working direction of wing plant motion. The wing is displaced relative to its axis in the PNWA to the final switch of the reverse 32, which switches on periodically trigger 15, and the wing for short time goes into the braking mode. Wing shape automatically is changed abruptly in the zones of the braking moment. Commands of control with a calculated delay are applied to the servo motors of flaps of corresponding wings.

The method of the kinetic energy transforming of the flowing environment in the rotational motion of the wings [66]. The method differs from the described above one that it contains criterion of optimization which is directly realized rotation moment of each wing due to the replacement of extreme pressure regulator of each wing 8 on its axis 7 for the extreme regulator of its main shaft turns 2 (Fig.3.28).

In the method, according to the invention, each wing 8 under the action of airflow rotates on the circular orbit and the vibrational motion around own axis 7. When received signal at the output of the rotation sensor of the main shaft, which does not exceed the nominal value U_n, extreme regulator of main shaft rotation on the basis of the controller operates angles of attack α of the control wing 8 relatively to the vector of resultant flow at all points of the circular orbit

Rotation in Fig.3.27, with the exception of zones for change moment sign and shape of the wing 13, acting through it servo of the flap 9 on the position of the control wing. At the same time, the controller memorizes commands, served on the servo motor of the control wing flap at all points of the circular orbit on each of its turns (0 - 2π), and calculates the lag when rotating along circular orbits of the remaining wings shifted relatively to control wing by $2\pi n / m$ (where m is the number of wings, N≥1-numbers of the remaining wings) taking into account the s signals of rotation sensor, and then transmits this data, as commands of control directly on the flap servos 9 of remaining wings 8. With the rotation of each wing, its axis 7 passes through the PNWA in Fig.3.26 and is shifted within of the housing 35 of the PNWA together with the bearing 34. In this case, the torque is formed, the maximum possible at a given airspeed and load on the main shaft 2.

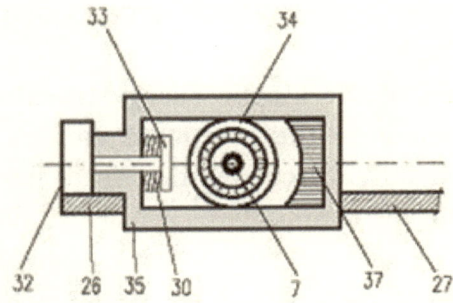

Fig.3.28. The design of the PNWA: 7-wing axis, 26- front B.L, 27- rear B.L, 30 - spring, 31- operational stop of PNWA, 32-limit switch reverse, 33 - pusher, 34 - bearing, 35 - body PNWA.

In the zones of change in the shape of the wings, the axis of the wings 7 in PNWA are shifted towards the reverse switch, periodically closing its contact, which gives a signal for switching on the trigger 15, alternately pushing the upper or lower mushroom-shaped stem and changing in these zones, the configuration of the wings symmetrically to a base sheet 26, 27. In this case, the nature of the mode of input into operating mode with the axes of the wings is shifted towards the working stop PNWA 31.

When the wing moves along a circular trajectory in zones 180 degrees + and 360 degrees + relatively resultant flow in the zone of the wing shape change the braking moment is formed, counteracting to working direction of the main shaft 2. And in vertical-axial wind installations of the axis of the wings from above and from below are rigidly fixed to the support disks 6, 12, which is perpendicular to the main shaft and rigidly connected with it. The wings' axes are fixed on the radii of these disks and the angles between these radii are $\beta = 2\pi/m$, where m- number of wings. To improve environmental characteristics of the outer circumference of the reference discs acellular protective net is attached.

The wind engine in Fig.3.29 is considered as an intermediate version of the optimization of the wind wheel, improvement of its design and system of management.

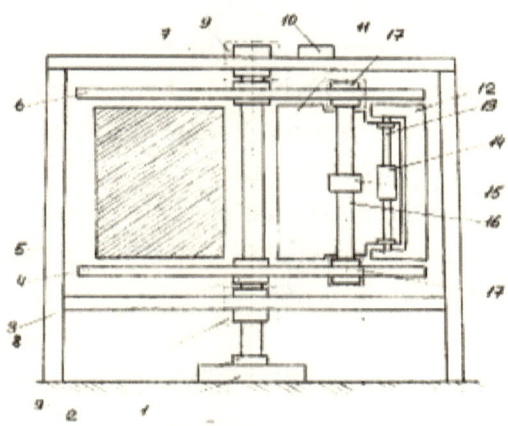

Fig.3.29. Wind turbine: 1 - block electromechanical converter, 2 - sensor rotation of the main shaft, 3 - frame, 4, 6 - support discs, 7-main shaft, 8-nut, 9 -support, 10 - flow velocity sensor, 11-blade, 12 - flap, 13 - axis of flap, 14 - node flap control, 15-node of determination of pressure, 16 - axis of blades, 11, 17-support.

The goal is to increase the take-off power from the unit of streamlined area, efficiency in the zone of a small speed of airflow (1.5 - 4.0 m/s) and low turnover associated with cheaper and higher reliability of the main shaft bearing units, reduced of parasitic noise and vibration – improving ecology. In addition, the placement of the bearing planes and their axes between disks increases stiffness, a reliability of construction, including through the creation of a regime control to protect against strong gusts of wind, and compliance with certain environmental requirements (the cellular net is fixed on the edges of the discs).

A specified goal is achieved by the fact that in the proposed construction of wind engine the principle of frontal resistance (Drag Principle) is used.

Block 1 of the electromechanical converter supports the optimal speed of rotation by regulation of the load of the generator that is included in its composition. The velocity of the moving blades at the circular orbit of rotation - ωR, determined by the speed sensor 2, must be 1/3 of the current air velocity, determined by the anemometer. In the developed wind engine operating mode is provided by the plane position of each of the blades 11, in which the maximum possible pressure on their own axis 16 is formed along the whole trajectory of circular motion. In the 1st and 2nd quadrants relatively the airflow vector of the blade is located perpendicular to the flow vector due to a control of flap position from the control unit of the flap 14 to the nominal pressure per axle. In the 3rd and 4th quadrants, blades are positioned at a certain angle relative to flow vector, providing some positive or zero moment when rotating in the working direction.

The velocity of rotation of the blades 11 along a circular orbit ωR (where R is the distance between the main shaft 7 and the axis blades 16) is determined by the speed sensor 2. The knot of control flap 14 (which includes the servo motor and its controller), operates in 2 modes.

Operating mode with extreme regulation. The criterion of quality, the indicator of extremum in our case is pressure on the wing axis 16. On extreme control the pressure of the blades 11 on their axis 16 does not exceed some calculated (specified) a nominal value of M_{nom} when exceeding this value of the pressure regulator of node 14 goes into stabilization mode and the pressure of the blades 11 on their axis is decreased. In the 1st and 2nd quadrants when moving along the circular orbit of the plane of the blades 11 are deviated from perpendicular to the vector of airflow (a streamlined area is decreased) and reduce pressure to M_n. At very high speeds of flow in the stabilization mode, the position of the blade planes 11 will approach in all quadrants to the position of the parallel flow vector (weathering).

Protection mode on the nominal pressure. Protection from emergency operation at high speed of airflow is provided by limiting the signal from the pressure sensor 26 to some nominal value M_{non} the axis 16 of the corresponding blade 11. When M_n is exceeded node 14 enters the mode stabilization. The pressure of the blades 11 on their axis decreases.

In the 1-st and 2-nd quadrants when moving in a circular orbit planes of the blades 11are deviated from the perpendicular position relative to the airflow vector (their streamlined area is decreased) and the pressure is decreased till M_n. At very high flow rates in stabilization mode of the unit 14 the position of the planes of the blades 11 will approach in all quadrants to the position parallel to the flow vector (weathering mode).

Fig.3.30 shows a top view of a simplified wind engine construction. Schematically depicted the position of one of the blades 11 with the flaps 12 in the process of their rotation in a circle in operating mode and the braking zone of blades by Δ degrees. In the 3rd and 4th quadrants at Zones 18, 19 positions of blades 11 are shown under the influence of flaps 12 (by commands from the flap control node 14).

Fig.3.31 shows the construction of the upper and lower supports 17 of each blade 11, which are adjacent to corresponding to the discs 4, 6. Fig.3.32 shows the node of pressure 15, its body 24 with a cover 25 is fixed to bearing plane 11 with a pressure sensor 26.

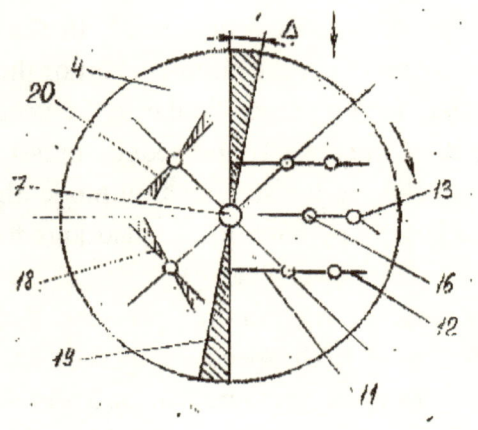

Fig.3.30. Top view of wind turbine design without upper disc:
4-lower disc, 7-main shaft, 11 - blades, 12-flaps, 13-axis flap,
16-axis blade, 18, 20 - zone work in the 3rd, 4th quadrants,
19 - zone transition.

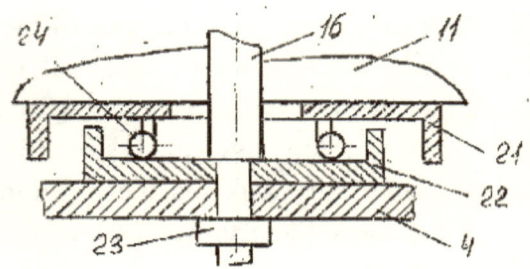

Fig.3.31. Construction of upper and lower supports: 4- lower disk,
11-blade, 16-axis blade, 21, 22-supporting elements, 24 -wheels.

This knot when the blade 11 moves relative to its axis 16 with the bearing 27 pushes it in the operating mode and decreases the pressure in the braking mode - in the zone Δ degrees to the value M0, Set by the spring 28. Moreover, the braking mode is short-lived in the zone Δ, and then under influence of the extreme regulator (node 14) and its Flap 12, the corresponding blade 11 again carries out pressure on its own axis.

In both modes, the pressure on axis 16 (with bearing 27) by means of a spring 28 and support plates 29, 30 acting on the sensor of pressure 26 is determined.

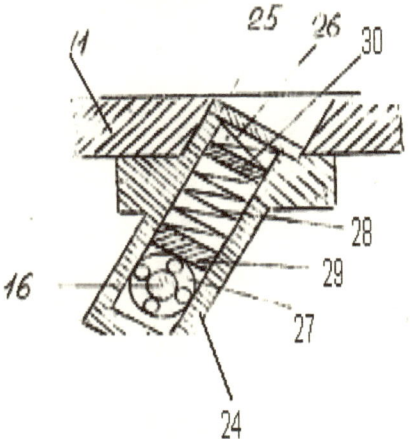

Fig.3.32. Pressure unit: 11-blade, 16-axis blade, 24-body, 25-cap, 26 - pressure sensor, 27- bearing, 28 - spring, 29, 30-support plates.

Below is a comparative analysis of developed wind engine efficiency in comparison with the known classic wind engine [67] of Fig.3.33, namely, of one blade with the rotation of its plane under variable angle α relative to the airflow in the 1st and 2nd quadrants for both wind engines at the same flow rate. The moment produced by the blade of the wind turbine

$$M_1 = h_1 F_1, \tag{3.4}$$

where the arm $h_1=(L_1/2)sin\alpha$, wherein L_1 – length of blade, $F_1=kS_1sin\alpha$, wherein S – area of blade.

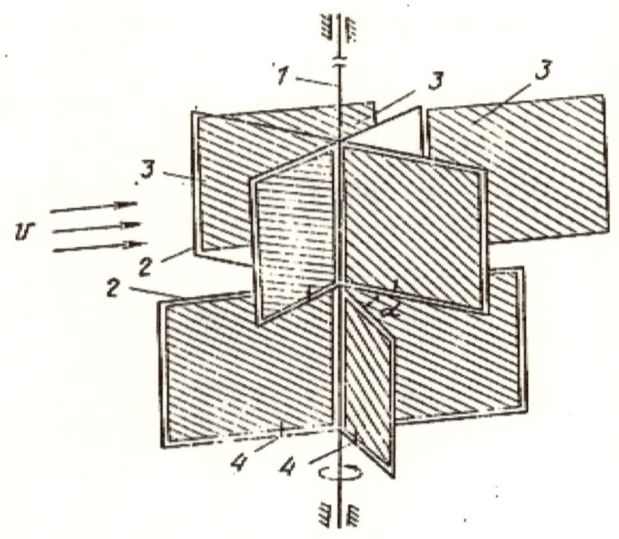

Fig.3.33. Wind engine: 1-shaft, 2 - bearing element, 3 - blade, 4 -supports, V- airflow.

The moment created by the wind turbine

$$M_2 = h_2F_2, \qquad (3.5)$$

where h_2 - the distance between the main shaft 7 and the axis of blade 16, let it be equal to $L_1/2$, $F_2 = kS_2$ and let $S_1 = S_2$.

The energy received from one blade on changing angle α in the range 0-180 degrees (1st and 2nd quadrants) for known wind engine and designed one is equal to the work of moments of the applied forces per rotation by an angle π

$$A_{\pi1} = (L_1/2)kS_1\int_0^\pi sin^2\alpha d\alpha = (L_1/2)kS_1 \pi/2 = \pi/4\, L_1kS_1, \quad (3.6)$$

$$A_{\pi2} = L_1/2\ kS_1\ \int_0^\pi sin\alpha\, d\alpha = 2\, L_1/2\ kS_1 = L_1kS_1. \qquad (3.7)$$

Thus, the proposed wind engine allows you to increase the efficiency of energy conversion of airflow with the same streamlined area. At developed wind engine it is foreseen to protect its construction from high-speed flows, improved performance on ecology: low speed and minimization of interference due to significantly reduce vibration and interference, and protective net from debris, birds, etc.

Easier way has been proposed to optimize when implementing the task considered above. The idea of optimization is reflected in Fig.3.34.

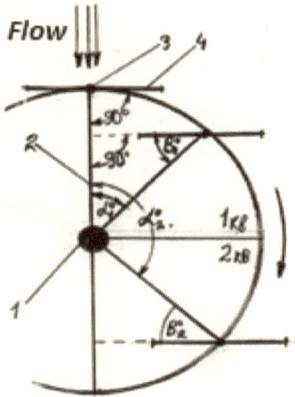

Fig.3.34. A variant of optimization of carousel wind wheel:
1-shaft with the control unit and position sensors, 2- carrier rod,
3- electric drive of the blade with a position sensor, 4 - blade.

When controlling the direction of the airflow, measurements with the angle sensors between the position of bars 5 and flow vector α_{11}, α_{21} by means of a simple calculator in the control node 1 determine the angles β_1, β_2, which are fulfilled by the electric drive of each blade 4. In this case, the relations are valid

$$\beta_1 = 90^\circ - \alpha_{11} \quad \text{и} \quad \beta_2 = \alpha_{21} - 90^\circ \qquad (3.7)$$

When you implement it, you do not need flaps and a node pressure, although electric actuators 3 with position sensors and a common control unit with a position sensor of the shaft 1, also associated with a weathervane are required.

The method of transforming the kinetic energy of airflow into the rotational motion of flat blade and wind energy plant that realizes it according to [68]. The proposed method provides high energy efficiency of the developed carousel wind plants. In the device of Fig.3.35, a block of electromechanical converter based on the generator is connected to the main shaft.

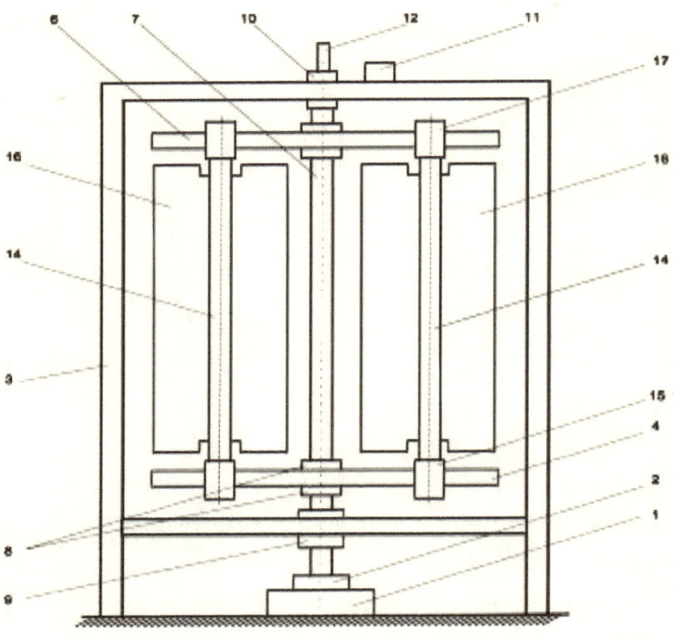

Fig.3.35. Carousel wind farm: 1- energy- converter, 2-shaft speed sensor, 3- frame,4, 6-discs, 5-support shafts, 7-shaft, 8-nut, 9-lower frame support, 10-shaft position sensor, 11-gauge of flow velocity, 12-weather vane, 13-flat blade, 14-axis blade, 15-controller with drive and the sensor of blade axis position.

Sensors of the rotational speed of the shaft and air velocity are connected to inputs of the converter and to its outputs of industrial network or group batteries of high capacity, with inverters, working on an industrial network are connected. Control Node Block generates such a load on the output of the generator, which provides a linear velocity ωR of the blades (where R is the distance between the shaft and the axis of the blade) approximately equal to 1/3 of the velocity of the airflow during movement of blades in a circular orbit. The angular position sensor of the shaft and wind vane are attached from above to the shaft, stator of the angular position sensor of the shaft is rigidly connected to the shaft, and its rotor is rigidly connected with the axis of the wind vane. Drive housing together with the controller are fixed on the lower disk, and drive, which includes a position sensor is rigidly connected to the axis of the blade. In connection with the movement of the axes of blades along a circular path, additional airflow vector acts on ach blade, tangent to the path of blades motion.

Due to low turnover of the carousel wind wheel and the presence of several blades effect of the additional flow is minimized. It is not taken into account in calculating the optimal position of the blades relative to the airflow. The flow entering carousel wind wheel is limited by upper and lower discs that are rigidly fixed to its shaft and increase the density of the stream flowing onto the blades. Slipping of the airflow from the top and bottom parts of the blade are missing.

Control of the position of the blades is carried out by driven with taking into account its own position sensor, the drive control command is generated by its controller on the basis of information - angle α, obtained from the sensor of angular position of the shaft relative to the air vector of flow, defined by a weathervane. With several blades of carrousel wind wheel, constant corrections are introduced in their controllers for the angle α-180, 120 or 90 degrees, etc., depending on the number of blades. The goal of the proposed method is to increase power, removed from a unit of streamlined area and efficiency in the zone of low air velocity – 1.5-4.0 m / sec.

Low-speed carousel wind wheel provides a cheaper construction and increases reliability of the main shaft bearing units and reduction of parasitic noises and vibrations. Fastening cellar net around the periphery of the disc circle, as well as above factors, increases environmental characteristics of the wind wheel.

Fig.3.36 illustrates the procedure for obtaining of the maximum moment $M = Fh$, where F is the force generated by flow on the blade, an h-the shoulder of this force relative to the shaft. The dependence between the rotation angle α of the shaft 7 relatively to the airflow vector and angle of plane β of the blade 16 relatives to the flow vector has been found.

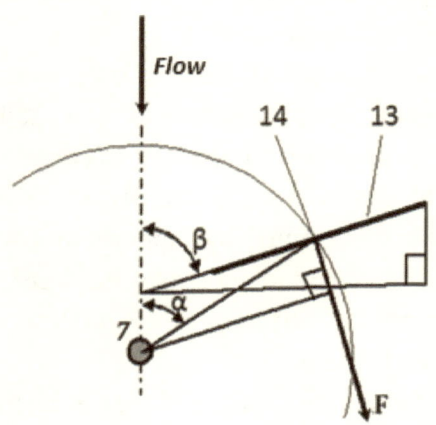

Fig.3.36. Force moment calculation:
7-shaft, 13-blade, 14 - blade axis.

The area of the projection of the blade on the plane perpendicular to the airflow vector is

$$S=S_0 sin\beta. \tag{3.8}$$

When optimizing the rotational speed of a carousel wind wheel, the speed of the axes of the blades $V_f = 1/3 V_{wf}$ and the force of pressure on the blade [10]

$$F = 2/9 \ C_X \rho \ S_0 \ sin\beta \ V^2{}_B \tag{3.9}$$

The moment is determined which the force develops according to (3.9) with respect to the shaft of the wind wheel. Shoulder h depends on the distance R between the shaft and the axis of the blade

$$h = R \ cos \ (\beta - \alpha). \tag{3.10}$$

Taking into account (3.9) and (3.10) the moment created by the blade has the following value at each point of the trajectory of blade motion

$$M_{\Pi} = F \ h = 2/9 \ C_X \rho \ S_0 \ sin\beta \ RV^2{}_B \ cos \ (\beta - \alpha). \tag{3.11}$$

It is necessary for each value of angle α on moving the blade to find the maximum value of the torque, which obviously depends on the angle β that determines position of the blade plane relative to the vector of airflow. The derivative of the moment M_π with β

$$M'_{\pi\beta} = \frac{2}{9} C_x \rho S_0 RV_B^2 \ [cos\beta \ cos(\beta - \alpha) - sin\beta sin(\beta - \alpha)] =$$

$$= \frac{2}{9} C_x \rho S_0 RV_B^2 \ cos(2\beta - \alpha) = 0. \tag{3.12}$$

Accordingly, $cos(2\beta - \alpha) = 0$, $2\beta - \alpha = 90^0$, and it turns ehe main analytical expression for the developed wind wheel

$$\beta = \alpha/2 + 45^0. \tag{3.13}$$

The relation between the angles β and α by (3.13) in the wind wheel provides the maximum value of moment (3.11) at any point of the trajectory of blade motion.

The obtained result corresponds to the presented in Chapter 2, the solution based on the optimization model of orthogonal wind turbine, for ideal conditions of the flat blade with substantially low turnover speeds.

The method of transforming the kinetic energy of airflow into the rotational motion of a flat blade [69]. Unlike the previous method, which designed for low airflow velocities (2-5 M/s) and low revolutions of the main shaft 2, the proposed method is designed for wind turbines working in zones with high airflow rates (up to 15 m/s) and higher speeds of the main shaft. In spite of coincidence of structures, materials, individual nodes, sensors, and control unit - a functional regulator based on the controller implements the more complex algorithm of position control of blades. The shaft 14 of each blade 16 is connected coaxially with the shaft of its own electric drive, turning the corresponding blade by the command from the controller.

The regulator conducts a continuous analytical calculation of the torque M_1 and the angle β to the control blade relatively to the direction of the airflow in each point of the circular trajectory of its shaft 14 according to information obtained from flow direction sensors, speed flow, rotation of the main shaft 2, angular position of shaft 14 with respect to the flow direction by electric drive with position sensor on the shaft of the control blade. The optimization method for calculation of maximum possible moment in an orthogonal turbine is used.

In counting system shown in Fig.3.37 the calculated moment of the flow forces applied to the blade, represented in the basic optimization model by formulas (2.37) and (2.38), with respect to a plane ideal blade (s_2=0; s_3=0) is transformed to the form

$$M_1 = Fh = \{F_0[\,sin\beta - kcos(\beta - \alpha)]\sqrt{1 + k^2 - 2k\,sin\,\alpha}\}Rcos(\beta - \alpha).$$

$$(3.14)$$

Investigation of the expression (3.14) on the extremum gives an opportunity to identify the optimal orientation of control blades of the wind wheel, that is, to determine optimal ratio $\beta=\beta(\alpha)$, ensuring maximum torque value M_1 of each blade 16 with the given air velocity $\overline{Vo}$.

The extremum condition

$$\frac{dM_1}{d\beta} = (sin\beta \, cos \, (\beta - \alpha) - k \, cos2(\beta - \alpha)) = 0 \qquad (3.15)$$

After the transformations, it becomes

$$\beta_K = (\alpha + arctg[k \, cos\alpha/(1-k \, sin\alpha)] + \pi/2)/2 \qquad (3.16)$$

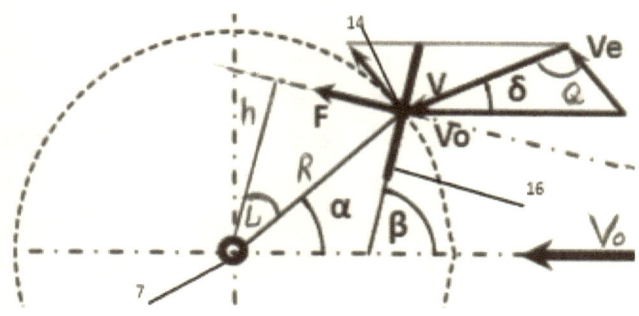

Fig.3.37. Method for calculating the torque of the wind wheel:
7-main shaft, 14-blade shaft, 16 - blade.

At the initial time, the wind wheel is motionless, however, the regulator generates the command β_k to electric drive with position sensor 15 of the control blade 16 according to (3.16), using information from sensors: flow direction, flow speed, rotation of the main shaft 7, angular position of the main shaft 7 with respect to the direction of airflow. The controller stores the angle values $B_k(t)$ of the control blade within each complete turnover ($0 < \alpha < 2\pi$), and for the orientation of the remaining blades constant corrections are present in controller to angles β_k-180, 120 or 90 degrees, etc., depending on the number of blades in the wind wheel, and the controller forms commands β_i for electric drives 15 of the remaining blades.

In addition, the obtained analytical expressions of torque on the shaft of the control blade allows on exceeding the nominal torque of M_{nom} at the exit of the regulator ($M_1 > M_{nom}$) to form commands for calculated angles β_k decrease and the delay β_i for their electric drives and reducing the torque on the shafts of all blades to the value of M_{nom} and with storm currents at exceeding the permissible speed of airflow - V_{max} by the command of the flow velocity sensor, the controller reduces β_k and β_i, the positions of all blades with respect to the direction of flow up to zero angles.

The method of channeling the energy of airflow by Fig.3.38 for wind power plants s of small and of average power. When it is implemented, it is necessary to control (regulate) the speed of turbine (of any type)　revolutions, and in some cases, regardless of velocity of airflow, and also to have in stock a powerful and capacious battery [9].

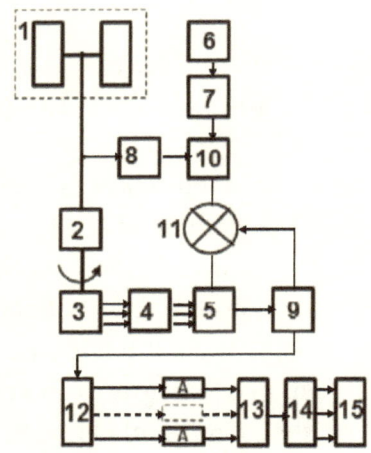

Fig.3.38. Wind power system of wind energy: 1-turbine, 2-reducer, 3-synchronous generator, 4- step-up transformer, 5- current regulator, 6-anemometer, 7-nonlinearity, 8-speed sensor,9-current sensor, 10-optimizer, 11-comparison node, 12 - charging current compensation throttle, 13-common bus batteries, 14-inverter, 15-industrial network.

The synchronous generator is connected via a reducer to the shaft of the turbine. Synchronous generator output via step-up transformer is connected to the current regulator of charge battery. At the output of the current controller N parallel battery circuits are connected to the nodes equalizing the charge currents at the input of each circuit (Leveling choke).

Operation of the full charge current channel (load) starts on voltage exceeding at the output of regulator over the rated voltage A (U_H), and up to this moment, the turbine spins at idling speed. The energy of the airflow is proportional to the flow velocity cubed V^3_{af}.

Due to the fact that voltage at the battery is changed a little charge current is set for a regulator of current by means of its proportional changing of ($KV^3_{af} > U_{nom}$), to obtain the initial signal from the anemometer output through nonlinearity. A nominal inverter is connected to the battery with work on special load or industrial network.

Using the wind farm directly from the network allows to obtain optimum turbine revolutions at the expense of setting the optimal load to the generator, varying coefficient K, that is, it allows to maintain an optimal speed of turbine rotation in wide speed range. Optimal speed of rotation assumes both effective energy, and favorable, safe regime for the design of the turbine - avoiding dangerous speeds, zones resonant vibrations, and the like. At high flow speeds when the total charging current exceeds the rated current of the generator, the command is given to change the angle of wing attack up to negative values and decrease Lift Force up to zero.

Final summary

When forming wind energy complexes the main task is providing an acceptable level of economic efficiency, implemented on the basis of a number of private technical solutions of problems. On the one hand, it is necessary to ensure energy efficiency of devices including maximum extraction of energy from the air stream, its transformation into electrical energy, accumulation of energy, its distribution and so on. On the other hand, the cost of manufacture and operation of wind stations should be economically acceptable. Accordingly, priority development trends of modern wind energy are directed, firstly, to the growth of capacity both single wind power plants, and their complexes, and secondly, on an improvement of aerodynamic characteristics of wind engines, and finally, the creation of effective regulation systems.

Almost the entire modern fleet of means of wind power generation consists of lobe (wing-bladed) wind plants. Collinear designs are recognized the most corresponding ones to the above requirements of efficiency among them occupying more than 90% of this sector. However, the most widespread propeller collinear plants according to expert estimates are close to the limit of the single capacity (10 MW).The most likely development trend of collinear installations is considered high-speed wind turbines with aerodynamic flow concentrators, as well as with adjustable aerodynamics of working blades. Such turbines are more compact in comparison with propeller devices (with comparable energy efficiency), and give additional scalability, as in single, and in multi-modular versions.

Significant interest is the application of vertical-collinear arrangement of wind plants which have advantages, such as comparative simplicity and reliability of the design in connection with the dynamic stability, ease of maintenance, additional scalability. Using of aerodynamic flow concentrators including vortex ones allows substantially to reduce dimensions of wind turbines while maintaining high performance of energy efficiency.

Considering a wide class of orthogonal wind plants, it is necessary to single out devices with unregulated and self-regulating wind engines. These are carousel wind turbines, sometimes equipped with special means of self-regulation - various elastic elements and other mechanical devices optimizing the interaction of working organs with airflow. Although these devices have low power and low energy efficiency their main advantage - the simplicity of construction and relative cheapness allows using such small wind power in hard-to-reach localities, in the field, etc.

Orthogonal vertical-axial wind plants with optimal control systems have significant prospects for effective large-scale applications. Their layout, especially in sea-based conditions on floating platforms in places of intense winds, gives significant opportunities for capacity expansion (up to 100 MW), with high dynamic stability and energy efficiency. Vane blades linear-convex section or sail-type device are used as regulated working bodies.

Rotation moment on the main axis of the wind engine and rotation speed of wind turbine is most often used as target parameters of regulation wind plants. Control actions are carried out through autonomous drives of the working parts of the wind engine, by means of changing their orientation - both in general and individual segments. Using of such complex drives is advisable to install a large unit power. To control the "smart" drive unit, processors that perform regulatory actions based on optimization calculations based on the developed methods are also used.

Keeping in mind economic feasibility and considerable increasing interest in wind energy, examined developments, taking into account the current trends, can find wide application in the national economy and power systems for various purposes - industrial, household and others.

References

1. Vozobnovlaemie istochniki energii. Teoreticheskie osnovi, tecknologicheskie, tehnicheskie harakteristiki, economika [in Russian] (Renewable energy sources. Theoretical bases, technologies, technical characteristics, economics). Z.A.Stychynsky, N.I.Voropay. Otto-von-Guerick-Universitat, Magdeburg, 2010, - 203 p.

2. Gordon V.V., Gubin V.E., Matveev A.S. Netradicionnie vozobnovkaemie energii. Izdatelstvo Tomskogo politechnicheskogo universiteta. [in Russian] (Non-traditional renewable energy sources. Publishing house of Tomsk Polytechnic University). 2009. - 294 p.

3. Krivtsov V.S., Oleinikov A.M., Yakovlev A.I. Neischerpaemaia energia. Kniga 1. Vetroenergogeneratori. Harkov: "HAI" [in Russian] (Inexhaustible energy: Book 1. Wind power generators. Kharkov: "KhAI"). 2003.-382 p.

4. Netradicionnie i vozobnovlaemie istochniki energii, kurs lekcii: uchebnoe posobie/ sost. V.A.Ageyev [in Russian] (Non-traditional and renewable sources of energy, course of lectures): textbook / comp. - Saransk, 2014. - 184 p.

5. Vetroenergetika. Pod redakciei D. De Renzo. [in Russian] (Wind power. Edited by D. de Renzo). Moscow: Energoatomizdat. 1982. – 271 p.

6. Sokolovsky Yu.B. Sovremennie vetroenergeticheskie ustanovki (obzor)   Elektrotehnika: setevoi elektroni nauchni jurnal. / Yu.B.Sokolovsky, A.Yu.Sokolovsky. [in Russian] (Modern wind power plants, review). Russian Internet Journal of Electrical Engineering, vol. 2, № 4, 2015. - p.27-38

7. Sokolovsky Y.B., Sokolovsky A.Y. Technical Proposals for Wind Turbine Structures. Journal SCIENTIFIC ISRAEL Technological Advantages, vol. 15, № 3, 2013.

8. Sokolovsky Yu.B. Ispolzovanie vetra – ekologicheski chistogo isrochnika energii/ Vestnik doma uchenih – Haifa [in

Russian] (The use of wind - an environmentally friendly source of energy / Bulletin of the House of Scientists. – Haifa): Volume XXIV, 2011. - p.16-21.

9. Sokolovsky Yu.B. Vetrovie energeticheskie ustanovki/ Vestnik doma uchenih – Haifa [in Russian] (Wind power plants. / Bulletin of the House of Scientists. – Haifa): volume XXXII, 2014. - p. 80-87.

10. Rozin M.N. Teoria parusnih ustanovok [in Russian] (Theory of sailing rigs) / http://www.rosinmn.ru/vetro/teorija_parusa/teorija_parusa.htm (reference date 20.04. 2016).

11. Mkhitaryan A.M. Aerodinamika: uchebnik [in Russian] (Aerodynamics: a textbook). - 2nd ed., remade and additioned.-M .: EKOLIT, 2012. - 448 p.

12. Bychkov N.M. Vetrodvigatel c effektom Magnusa: rezultati modelnih issledovani / Teplofisika I aeromehanika [in Russian] (Wind turbine with the Magnus effect: Results of model studies / Thermophysics and Aeromechanics), vol. 11, No. 4, 2004. - p. 583-596.

13. Avtokolebatelnie sistemi / Enciklpedia po mashinostroeniu [in Russian] (Self-oscillation systems / Encyclopedia on Engineering) – URL: http://mash-xxl.info/info/12422/ (reference date 21.12.2016).

14. Weinberg V.Ya. Energopreobrazovatel [in Russian] (Power converter). / Patent No. 2253747, published on June 10, 2005.

15. R.A.Melnikov. Alternativnii kilovatt [in Russian] (Alternative kilowatt). - URL: http://www.electroveter.ru/ (reference date 17.10.2016).

16. Energeticheskaia kompania "Energia Disisn" [in Russian] (Energy Company "Energy design"). URL: http://e-ds.promportal.su/groups/27421/vetrogeneratori (reference date 07.12. 2015).

17. VEU s vertikalnoi ociu vrashenia: cravnitelnaia ocenka tehnicheskih reshenii I perspektivi razvitia [in Russian] (Wind turbine with a vertical axis of rotation: comparative evaluation of technical solutions and prospects of development). URL: http://www.energyland. info/analitic-show-52412 (reference date 07.12.2015).

18. Vetrogenerator s vertikalnoi ociu vrashenia: proshloe, nastoiashee, budushee [in Russian] (Wind turbine with vertical axis of rotation: past, present, future). URL: http://www.itsintez.com/ about/press/ publications/publications_ 51.html (reference date 10.06.2016).

19. P.Belyakov. Obzor rotornih vertikalnih vetroustanovok. (Overview of rotary vertical wind plants.). URL: nergyfuture.ru/prekrasnyj-obzor-rotornyx-vertikalnyx-vetroustanovok-ot-uchenyx-ix-voronezha (reference date 12.09. 2016).

20. Rotor Darie. (Darrieus Rotor) URL: https://ru. wikipedia.org (reference date 22.09. 2016).

21. Grahov Yu.V. Rotor vetranoi ustanovki s vertikalnoi osiu [in Russian] (Rotor wind turbine with Vertical axis) / Yu.V.Grahov, V.P.Krivospitsky, V.F.Maksimov, E.V.Solomin, Richard Halstead, Glen Dahlbakka. Patent RU 2347104, C2, F03D 3/06. The patent holder of OOO GRTS - Vertical. Published on 20.02.2009.

22. Kyan V.P. Optimizacia rabochih harakteristik polnomashtabnogo maketa vetrorotora Darie s pramimi upravlaemimi lopastami / Prikladna gidromehanika [in Russian] (Performance optimization Full-scale mock-up of the Darrieus wind farm with direct Controlled blades / V.P.Kayan, A.G.Lebed / Applied Gidromekhanika), vol.12, 2010. - p. 26-35.

23. Kayan V.P., Dovgy S.O., Boyko P.M., Lebed O.G. Vitrosilova ustanovka [in Russian] (Wind power plant) / Patent

of Ukraine No. 16097 for useful model, IPC F03D 3/00, F03D, 7.06.2006. - 174 p.

24. Rotor Savoniusa [in Russian] (The Savonius Rotor). URL: http://www.wetroenergetika.ru14%20savoinus.html (reference date 01.10. 2016).

25. Soloviev A.P. Rotor tipa Saviniusa s polimi lopastami [in Russian] (Rotor of Savonius type with hollow blades). / Patent of the Russian Federation RU No. 2101557, the date of the commencement of the patent on 28.07.1994.

26. A.I.Mozgovoy. Karuselnoie vetrokoleso i lopast karuselnogo vetrokolesa [in Russian] (Carousel wind wheel and blade carrousel wind wheel). / Patent RU No. 2202048, was published on February 17, 1998.

27. Alternativnie istochniki energii [in Russian] (Alternative energy sources). URL: http://www. artadmirescom/www/erilon2/energy/veter/types/ (reference date 01.10. 2016).

28. Innovacionnaia vetranaia turbina [in Russian] (Innovative wind turbine) IMPLUX, URL: http://gizmod. ru/2011/05/19/ innovacionnaya- vetryanaya- turbina- implux/ (reference date 02.06. 2016).

29. Rossiiski vetrogenerator s povishennim KPD [in Russian] (Russian wind generator with increased efficiency). URL:http://www.computerra.ru/125648/rossiyskiy-vetrogenerator -s-ovyishennyim-kpd/ (reference date 25.10.2016).

30. Soloviev A.P. Sostavnoi rotot tipa Magnusa [in Russian] (Magnus type compound rotor). / Patent RU No. 2213883 F3D3 / 00, published October 10, 2003.

31. Osnovnye vidy vetrogeneratorov i ih harakteristiki [in Russian] (The main types of wind turbines and their characteristics). URL: http://genport.ru/article/osnovnye-vidy-vetrogeneratorov-i-ih-harakteristiki (reference date 30.04.2016).

32. Vetroelektrostancia s koleblushimsia vertikalnim cilindrom [in Russian] (Wind power station with oscillating vertical cylinder). URL: http://www.vortexosc.com/modules. php?name=Content&pa=showpage&pid=173 (reference date 12.07.2016).

33. Novi proekt vetrofermi bez vetrakov [in Russian] (A new wind farm project without wind turbines). Russian Internet Journal of Electrical Engineering, vol. 2, №4, 2015. URL: www.electrical-engineering.ru (reference date 30.02.2016).

34. R.A.Serebryakov, A.B.Kalinichenko. Vihrevaia vetroenergetika [in Russian] (Vortex wind energetics). – URL: http://www.sovstroymat.ru/2001_11_14.php (reference date 23.05. 2016).

35. Enercon E-126. Sami Bolshoi vetrogenerator v mire [in Russian] (The largest wind generator in the World). URL: http://grandstroy. blogspot.co.il/2012/03/ enercon-e-126.html (reference date 05.07. 2016).

36. Sposob povishenia effektivnosti vetrogeneratorov [in Russian] (A method for increasing of wind generators efficiency). - URL: https://vseonauke.com/ 383712280821631828/ najden-sposob-povysheniya-effektivnosti -vetrogeneratorov/ (reference date 05.07.2016).

37. Seven ways floating turbines. URL: http://www. windpowerengineering.com/construction/seven-ways-floating-turbines-solve-offshore-conundrums/ (reference date 05.07. 2016).

38. Aerogenerator turbine sets sail for a greener future. URL: https://www.theguardian.com/technology/2008/jan/29/ wind. energy.aerogenerator (reference date 05.07.2016).

39. Khaskin L.A. Bashna iz vetroenergeticheskih modulei [in Russian] (Tower of wind energy modules) - Moscow: Nauka i zhizn, No. 9, 2003. - p. 70-72.

40. Constant Power Control and Fault-Ride-Through En hancement of DFIG Wind Turbines with Energy Storage / Department of Electrical Engineering University of Nebraska Lincoln 223N Walter Scott Engineering Center Lincoln, NE 68588-0511 USA. - p. 1-11.

41. Rudion K., Orths A., Styczynski Z.: Modelling of Variable Velocity Wind Turbines with Pitch Control. Proceedings of the 2-nd International Conference on Critical Infrastructures, 10.2004, Grenoble, France. - p.25-27.

42. Control of DFIG Wind Turbine With Direct-Current Vector Control Conguration. / Shuhui Li, Senior Member, IEEE, Timothy A. Haskew, Senior Member, IEEE, Keith A. Williams, and Richard P. Swatloski. - p.359-367.

43. Turysheva A.V. Analiz effectivnosti primenenia sovremennih elektrochimicheskih condensatorov/ Sovremennaia tehnika I tehnologii [in Russian] (Analysis of effective using of modern electrochemical capacitors / Modern technology and technology) - № 5, 2014. URL: http://technology. snauka. ru/2014/05/3856 (reference date 12.06. 2016).

44. Zhukovsky N.E. Vetranaia melnica tipa NEJ [in Russian] (Windmill of type NEJ) (1920 y). - URL: http://rosinmn.ru/VETRO_Ideal_windmill.htm (reference date 09.09. 2016).

45. Rozin M.N. Teoria idealnogo vetraka [in Russian] (Theory of the perfect windmill) / - URL: http:// www.rosinmn.ru/vetro/teorija_idealnogo_vertjaka/ shablon.htm, 04.08.2008. (reference date 20.09.2016).

46. Okulov V.L., Sorensen Zh.N. Idealni vetrak s konechnim chislom lopastei/ Institut im. Kutateladze Sibirskogo otd. Rossiiskoi Akademii nauk [in Russian] (The perfect windmill with a finite number of blades. The Institute. S.S. Kutateladze Siberian Branch of the Russian Academy of Sciences),

Novosibirsk // Reports of the Academy of Sciences, Vol. 420, No. 4, 2008. - 428-483 p.

47. V.Rotkin, Y.Sokolovsky. Energy efficiency of bladed wind turbine. Optimization model. Journal SCIENTIFIC ISRAEL Technological Advantages, vol.18, № 2, 2016. - p. 111-122.

48. Gurevich V.A., Sokolovsky Yu.B., Sokolovsky A.Yu., Frolov E.A. Vetrosilovaia energoustanovka vrachauchegosa tipa [in Russian] (Wind power plant rotating type). / Patent RU No. 2563558. published on 20/09/2015.

49. Romanov G.A. Cilindricheskaia vetroturbina [in Russian] (Cylindrical wind turbine). Patent RU No. 2156885. published 27.09.2000.

50. Sokolovsky Yu.B., Sokolovsky A.Yu., Ivanova O.Yu., Sokolovsky D.Yu. Ekologichnaia i energoeffektivnaia vetroturbina na gorizontalnom valu [in Russian] (Eco-friendly and energy efficient wind turbine on a horizontal shaft). RF Application No. 2016136144. Date of receipt of the application 08.09.2016.

51. Sposob orientacii vetroustanovki [in Russian] (Method for orienting of wind plant) Anhui Hummer Dynamo Co. Ltd. / URL: http://hummer.china-manufacturer-directory.org/windmill-generators/2000w_Windmill_Turbine/ (reference date 22.05.2016).

52. Gurevich V.A., Sokolovsky Yu.B., Sokolovsky A.Yu., Kheifets A.B. Sposob orientacii vetroenergeticheakih ustanovok s gorizontalno-osevimi propellernimi turbinami [in Russian] (The way of orientation of wind energy plants with horizontally-axial propeller turbines). / Patent of the Russian Federation No. 2588914. Date of registration 07.06.2016.

53. Seleznev N.V. Vetroelektrostancia [in Russian] (Wind power station). / Patent RU No. 2285147 C1, F03D03 / 00, published on 10.10.2006.

54. Murashevsky V.V. Vetrovaia energeticheskaia ustanovka [in Russian] (Wind power plant). / Patent RU No. 2277642, published 10.06.2006.

55. Sokolovsky Yu.B., Gurevich V.A. Vetrovaia energeticheskaia ustanovka [in Russian] (Wind power plant). / Patent RU 2484296 C2, published on 03.08.2011.

56. Gurevich V.A., Sokolovsky Yu.B. Karuselnoie vetrokoleso [in Russian] (Carousel Wind wheel). / Patent RU No. 2498109, published 10.11.2013.

57. Gurevich V.A., Sokolovsky Yu.B., Sokolovsky A.Yu., Frolov E.A. Turbina dla ispolzovania kineticheskoi energii vetra ili potoka vodi. [in Russian] (Turbine for kinetic energy using of wind or water flow). / Application: No. 2014104114, date of receipt 05.02.2014.

58. Sokolovsky Yu.B, Sokolovsky A.Yu., Frolov E.A. Karuselnoie vetrokoleso [in Russian] (Carousel wind wheel). / Application of the Russian Federation No. 2016108479, date of registration 09.03.2016.

59. J.Sokolovsky, A.Heifetz, V.Sosenushkina. Wind Power in Israel with the Use of Sea. Coasters: Technical and Economic Calculation in the Project Business Plan. Journal SCIENTIFIC ISRAEL Technological Advantages. vol. 15, № 2, 2013, - p. 112-113.

60. Y.B. Sokolovsky., A.Y.Sokolovsky. Technical Proposals for Wind Turbine Structures. Journal SCIENTIFIC ISRAEL Technological Advantages. vol. 15, № 3, 2013. - p.19-21.

61. Sokolovsky Yu.B., Sokolovsky A.Yu., Limonov L.G. Povishenie effektivnosti vetrovih energeticheskih ustanovok. Ukraina: jurnal "Energosberejenie, energetika, energoaudit" [in Russian] (Efficiency increasing of wind energy plants. Ukraine, magazine "Energy Saving, Energy, energy audit») №9 (127), 09.2014. - p. 28-37.

62. Sokolovsky Yu.B., Sokolovsky A.Yu., Limonov L.G. O primenenii vetrovih energeticheskih ustanovok/ Jurnal "Elektrotehnicheskie I kompiuternie sistemi", Odessa: Nauka i technika [in Russian] (On the application of wind energy plants. Magazine "Electrotechnical and computer System", Odessa: Science and Technology), No. 16 (92), 2014. p. 7 - 15.

63. V.Rotkin, Yu.Sokolovsky, A.Sokolovsky, E.Frolov. New variants of wind energy plants. Journal SCIENTIFIC ISRAEL - Technological Advantages, vol.18, № 4, 2016. - p. 89 - 99.

64. Bokay V.I. Sposob preobrazovania kineticheskoi energii tekuchei sredi vo vrachatelnoie dvijenie krila i ustanovka dla osuchestvlenia etogo sposoba [in Russian] (The method of transforming kinetic energy of the current medium into rotational motion of the wing and the plant to implement this method). / Patent of Russia No. 2157919, published on October 20, 2000.

65. Gurevich V.A., Sokolovsky Yu.B., Sokolovsky D.Yu., Frolov E.A. Sposob preobrazovania kineticheskoi energii potoka vo vrachatelnom dvijenii krila i ustanovka dla osuchestvlenia etogo sposoba [in Russian] (The method of transforming kinetic energy of the flow in rotational motion of wing and plant for the implementation of this method). Patent RU No. 2589569. Published on 07.10.2016. Bull. No. 19.

66. Sokolovsky Yu.B., Sokolovsky A.Yu., Ivanova O.Yu., Sokolovsky D.Yu. Sposob preobrazovania kineticheskoi energii tekuchei sredi vo vo vrachatelnoie dvijenie kriliev [in Russian] (The method kinetic energy conversion of the current medium into the rotational movement of the wings). / Application of the Russian Federation No. 2016118917, date receipts on 29.04.2016.

67. Golovko V.M., Makievskaya V.E. Vetrodvigatel [in Russian] (Wind turbine). / The author's certificate №1663226 A1, published 15.07.1991.

68. Gurevich V.A., Sokolovsky Yu.B., Sokolovsky A.Yu. Sposob preobrazovania kineticheskoi energii vozdushnogo potoka vo vrachatelnom dvijenii ploskoi lopasti [in Russian]   (The method of transforming kinetic energy of airflow in rotary motion flat blade). RF application No. 2016108449, the date of receipt 09.03.2016.

69. Sokolovsky Yu.B., Rotkin V.M., Ivanova O.Yu. Sposob preobrazovania kineticheskoi energii vozdushnogo potoka vo vrachatelnoe dvijenie ploskoi lopasti [in Russian]   (The method of converting the kinetic energy of airflow into rotational motion of a flat blade). / Application RF No. 2016125205, the date of receipt of the application on 24.06.2016

www.ingramcontent.com/pod-product-compliance
Lightning Source LLC
Chambersburg PA
CBHW032022170526
45157CB00002B/819